Being and the Cosmos

From Seeing to Indwelling

Robert E. Wood

THE CATHOLIC UNIVERSITY OF AMERICA PRESS
Washington, D.C.

To all my colleagues, past, present, and to come

The paper used in this publication meets the minimum requirements of American National Standards for Information Science—Permanence of Paper for Printed Library Materials, ANSI Z39.48-1984.
∞

Library of Congress Cataloging-in-Publication Data
Names: Wood, Robert E., 1934– author.
Title: Being and the cosmos : from seeing to indwelling / Robert E. Wood.
Description: Washington, D.C. : The Catholic University of America Press, 2018. |
Includes bibliographical references and index.
Identifiers: LCCN 2018021324 | ISBN 9780813231174 (pbk. : alk. paper)
Subjects: LCSH: Ontology. | Cosmology.
Classification: LCC BD331 .W85 2018 | DDC 111—dc23
LC record available at https://lccn.loc.gov/2018021324

Being and the Cosmos

Contents

Acknowledgments

Three of these chapters appeared in previous publications: "On Seeing: A Phenomenological Approach to Neuro-Psychology" appeared in *Science, Reason, and Religion*, Proceedings of the American Catholic Philosophical Association Convention (Fall 2011): 121–34.

"First Things First: On the Primacy of the Notion of Being" appeared in the *Review of Metaphysics* 67, no. 4 (June 2014): 719–41.

"The Cosmos Has an Inside" appeared in *Existentia* 26, nos. 3–4 (2016): 465–80.

Introduction

MY AIM IN THIS WORK is to reestablish a speculative view of the cosmos that goes back to the ancient Greeks and that corresponds to the holism of contemporary physics. There are two sets of problems in contemporary thought that militate against any such attempt. Most widespread is scientific reductionism in biology and neuroscience that explains awareness in terms of the mechanisms that underlie it. The second is the widespread attack in philosophy itself on speculative holism by deconstruction and anti-foundationalism. My tack against both is to make explicit the character of the mind that sees and thinks, that actively takes up commitment to the truth available in the disciplines involved. And the basic ground of my own position rests upon the functioning of the notion of Being that opens up the question of the character of the Whole and the human being's place in it. I thus position the treatment of the notion of Being as foundation and as orientation toward the Whole between the attack on reductionism and on deconstruction and anti-foundationalism. I conclude with a multidimensional sketch of an evolutionary view of the cosmos whose initial phases contain the potentialities for life, sensibility, and intellect as cosmic *telos*.[1] The holism of contemporary physics has to be reconfigured in terms of this observation. Matter itself has to be re-minded, and mind itself matters.[2]

THERE ARE THREE crucially important developments in modern science: in physics a speculative cosmology, in biology evolutionary theory, and in neurophysiology the progressive uncovering of the control systems in the brain and the nervous system it synthesizes. The development of the latter science is rightly called "the new frontier."

From its inception, modern scientific investigation followed along the lines laid down by Descartes of two separate regions, that of thought and that of extension. Though Descartes viewed the latter as a single whole, under the Newtonian view scientific theory assumed the mechanism of regular external interaction between discrete picturable units. Though Newton said he did not base his thought on hypotheses, his positing of gravity, a nonpicturable attraction between discrete bodies, was indeed a hypothesis.

In the twentieth century the underlying model shifted, along the lines of Descartes's view of extension as a single whole, to an overall space-time-energy matrix within which things are particular enfoldings. At the deepest level uncovered, one can no longer picture the elements that exhibit both particulate and

1. See Thomas Nagel's surprising *Mind and Cosmos: Why the Materialist Neo-Darwinian Conception of Nature Is Almost Certainly False* (Oxford: Oxford University Press, 2012). He argues that, because we can lay hold of *eternal truths* in logic and mathematics, we cannot be an accident of evolution, but are its goal.

2. See Galen Strawson, *Real Materialism and Other Essays* (Oxford: Clarendon, 2008). The book could equally well have been entitled *Real Spiritualism*, but the current milieu is better served by Strawson's choice.

wavelike symptoms; and in the larger picture, black holes are inferable, not picturable. But over time, especially through progress in neurophysiology, the mind, originally conceived as separate by Descartes, came to be reduced to the physical system so that at present, reductionism is the default mode for scientific and consequent popular thinking as well as for much of philosophic thought.

The great empirical developments followed the pattern of visual evidence provided by ever more complex microscopic, telescopic, and manipulative inventions that allowed for inference to nonpicturable features. My contention is that, though visual evidence leads us on pragmatically, in neurophysiology it systematically misleads us speculatively into thinking that all that is has the character of visual objectivity. What seeing reveals are actual individuals, immediately present: in the case of living beings, appearing as contained within their visible boundaries and with everything linked to their antecedents and consequents in a visually verifiable seamless connection. Taking that as the type of all evidence leads to what I call "the optomorphic fallacy."[3]

Physics has moved light-years beyond the seeable, but optomorphism still dogs biology and especially neuroscience. The latter is making spectacular progress in understanding how the nervous system works by simply bypassing the nature of awareness. But awareness still remains the basis for the advancement of science itself. Anticipating future developments in neuroscience, many thinkers in philosophy of mind attempt to develop the change in categories required by reductionism.

3. George Wilhelm Friedrich Hegel made a similar observation in his critique of *Vorstellung*, or "representation," sometimes translated as "picture thinking." *Vorstellung* is thinking in terms of separate items externally related, such as are presented in seeing; Hegel, *Encyclopaedia of Philosophical Sciences*, vol. 1, *The Encyclopaedia Logic*, trans. T. Gaerets, W. A. Suchting, and H. S. Harris (Indianapolis: Hackett, 1991), §20:50.

In contrast to what seeing presents, reflection upon seeing and upon reflection itself presents not simply individuals, but underlying powers, *universal* orientations correlated with natural *kinds*. Further, in the case of animal and human beings, awareness involves living *outside* their own visually discernible bodies and *beyond* the immediate through learning from the past and appetitive anticipation of the future. Humans live outside their privacy through a common language that presents sensory forms as stand-ins for the apprehension of the essential distinctions by what we have come to call "intellect," which thinks in terms of space and time as encompassing wholes.

The basis of intellectual activity, I will claim, lies in the notion of Being that, referring us to everything and to everything about everything, grounds the a priori forms of space and time as encompassing forms for that region of being, the material universe, to which we, as intellectually embodied beings, belong. In this way, the notion of Being grounds the capacity to recognize the universal and, by reason of the distance from all determinants in being referred to the Whole, the capacity to introduce novel lines of causality through free acts.

These considerations link up with physics, evolutionary biological theory, and neuroscience by installing the "inwardness" of awareness in an evolutionary view that deepens and grounds scientific theory. The mind of the scientist advances the understanding of its own antecedents by being included in the overall theory. Along Schellingian lines, matter has to be understood, not simply in terms of visual modeling and thus in terms of exteriority or as other than mind; it has to be understood as having, in its initial stage, the potentialities of life, awareness, and reflective awareness lying within it. Those potentialities do not lie in individual elements but in the system, in the particular sorts of relations that unlock them. The cosmos has an inside that is central to the development of the cosmos itself. It should be understood, not only in terms of the lowest discernible level of material causality, but in terms of the final causality of its end-state as implicit in the beginning. As Thomas Nagel observes, it requires the reintroduction of teleology as a cosmic principle: the human mind is the cosmos waking up to itself.

THE PRESENT WORK begins in chapter 1, "What Is Seeing?," an encounter with neurophysiology, by zeroing in on that which provides the wealth of evidence upon which scientific theory is built—namely, the act of seeing. The chapter explores the context within which vision typically operates by analyzing the field of awareness, sensory and intellectual, involved in the ophthalmologist examining her patient.

Chapter 2, "First Things First: On the Priority of the Notion of Being," develops the way in which the notion of Being operates in human experience, opening out the question of the Whole and providing the principle of noncontradiction for moving theoretically, by description and inference, into the Whole. It also explores the ways in which the founding *notion* of Being, as initially empty reference to the Whole, has led to various *concepts* of Being throughout Western philosophic history in Parmenides, Plato, Aristotle, Aquinas, Scotus, Spinoza, Hegel, Heidegger, and Derrida. This very brief history illustrates the differing directions in which the notion of Being has been developed conceptually.

If chapter 1 was an encounter with contemporary reductionism, chapter 3, "The Undeconstructible Foundations of Human Existence," is an encounter with contemporary deconstruction and anti-foundationalism. It follows the lines of evidence presented in chapter 1 and their grounding in the notion of Being in chapter 2. The basic contention is that the structure of the field of awareness and its operation with the notion of Being, along with the principle of noncontradiction cogiven with it, are the factual foundations for all human development and the basis for whatever deconstructive efforts we might make. This clears the ground for chapter 4, “The Cosmos Has an Inside.”

Here we take up the evidences we have marshaled and develop our approach through the use of the image of *dimensions*. Beginning with the four dimensions of the space-time universe, we have an instance of irreducible distinction with inseparable conjointness of these dimensions in whole entities. Taking the physicist’s distinction between four unfolded and ten “curled up” dimensions, we follow the latter into the interior dimensions of life, awareness, and reflective awareness, which, we maintain, are irreducibly distinct both from each other and from the bodies they inform, but are also inseparably conjoined thereto. Sentient and rational awareness involve, through their interiority, different ways of being in space and time than do nonliving things.

The depth of interior life is a matter of the heart, the residue of past experience that is the default mode for our choices. It sets up in each of us a unique magnetic field of attractions and repulsions. It is the ground of the way humans dwell in the world.

We push on ultimately to infer the “Within” of the first phases of the cosmos as a pool of potentialities, unlocked sequentially over immense periods of time through random mixing and leading to human existence. Humanness is there potentially at the Big Bang as the condition for the possibility of the future manifestation of the world-process itself and for the creative development of the human world within it.

These chapters were independently generated. Chapter 1, “What Is Seeing? A Phenomenological Introduction to Neuropsychology,” was delivered several times: at the University of Dallas, at the annual convention of the North Texas Philosophical Association, at the Dallas Philosopher’s Forum, at the University of Texas Southwestern Medical School, and at the annual convention of the American Catholic Philosophical Association. It appeared in the proceedings of the latter.[4] Chapter 2, "First Things First: On the Priority of the Notion of Being," was my first presidential address to the North Texas Philosophical Association in 2011, which appeared subsequently in the *Review of Metaphysics*.[5] Chapter 3, "The Undeconstructible Foundations of Human Existence," was delivered as my second presidential address to the same association in 2012. Chapter 4, "The Cosmos Has an Inside," was my third and final presidential address to that assembled association in 2013 and was published in *Existentia* in 2016.[6]

There is some overlap among the various chapters. But, since repetition better secures what is read (*Repetitio est mater studiorum*), I have decided to retain the repetitions.

Institute of Philosophic Studies
University of Dallas

4. Robert E. Wood, "On Seeing: A Phenomenological Introduction to Neuro-Psychology," *Science, Reason, and Religion*, Proceedings of the American Catholic Philosophical Association Convention (Fall 2011): 121–34.

5. Wood, "First Things First: On the Priority of the Notion of Being," *Review of Metaphysics* 67, no. 4 (June 2014): 719–41.

6. Wood, "The Cosmos Has an Inside," *Existentia* 26, nos. 3–4 (2016): 465–80.

1 What Is Seeing?

A Phenomenological Approach to Neuropsychology

What is most obvious and universal in experience is most frequently overlooked, probably because it is too familiar to evoke special comment.

—Errol Harris, *Nature, Mind and Modern Science*

Light-minded men "suppose in their simplicity that the most solid proofs about such matters are obtained by the sense of sight."

—Plato, *Timaeus*, 91e

WHAT IS SEEING? Physicists can trace the route of light radiating outward from light sources as it is partially absorbed and partially reflected off of bodies in the environment until it arrives at a seeing organism. Physiologists can investigate light's passage through the pupils and the lenses through which it impacts the rods and cones of the retina, setting off the electronic transmission of the impact via a set of electrochemical switches by means of which it ends up in the visual cortex at the back of the brain.

One typical approach to seeing occurred a few years back when Francis Crick (of Crick and Watson double-helical DNA fame) wrote a book called *The Astonishing Hypothesis: The Scientific Search for the Soul.*[1] The book zeroed in on seeing as the initial focus in a long-range project of examining the functioning of the brain to explain awareness through the "astonishing hypothesis" that eliminates the soul in favor of observable mechanisms. Crick presented a detailed but also somewhat scattered report on what was known about the process of physiological functioning with regard to seeing in conjunction with something of experimental psychology and suggestions for lines of research.

Crick rightly claimed that we had better know something about visual consciousness to explain it in terms of neural functioning, but what he brought forth attended only to isolated studies in experimental psychology such as Gestalt shifting or the binding of moments involved in minimal awareness, but not the total field of awareness within which visual presentation occurs. Further, what he presented did not draw upon the experimental psychological reports. And, in fact, his whole procedure is peppered with anthropomorphic descriptions of knowings, preferrings, and seekings on the part of the neurons. My claim is that what is needed first and foremost in the "search for the soul" is an inventory of the overall structure of the full field of awareness involved in human visual functioning.[2]

1. Francis Crick, *The Astonishing Hypothesis: The Scientific Search for the Soul* (New York: Simon and Schuster, 1994). Though this appeared several years ago, it demonstrates a common orientation.

2. In an extended confrontation with reductionists, P. M. S. Hacker and Maxwell Bennett, in *The Philosophical Foundations of Neuroscience* (Oxford:

From examining neural functioning, it is not uncommon—in fact, it is "natural" to conclude, as Crick does—that what we are seeing when we see are images in the visual cortex. (I will have something to say later about why I think the conclusion seems natural, even obvious.) If we granted the last claim, *not one thing* has been said about seeing except that, whatever it is, it takes place in the back of the cortex, and what are viewed are images resulting from the highly complex neurophysiological process. Whatever is said about the neurophysiological process is visually accessible, and reporting on that is good science. But when a scientist goes on to claim that what we see are images in the back of the brain, we have moved from science to philosophy, whose claims have different methods of verification. What then *is* seeing?

Let's come at it first from the essential features of what is seen. The generic object of seeing is color. Even the so-called color-blind see black, white, and gray, often (oxymoronically) called "achromatic colors." And color necessarily involves extension. One cannot even imagine a color that is not extended. In things, extension appears terminated by shape surrounded by the encompassing extension of the sky. But color is not a thing; it is rather a dependent feature of a given thing: the tree, the sky, the friend. Visible things move within the field of awareness, so the shape and motion of things are cogiven with the visual object.

When we open our eyes from a fixed position, a 180-degree arc opens up, revealing a panoply of things that remain margin-

---

Blackwell, 2003), set ordinary language use over against reductionist attempts. Their direct encounter with David Dennett and John Searle drew a huge crowd at a convention of the American Philosophical Association and was made available in print as P. M. S. Hacker, Maxwell Bennett, Daniel Dennett, and John Searle, *Neuroscience and Philosophy: Language, Mind, and Brain* (New York: Columbia University Press, 2007). The approach I am taking here gives the phenomenological underpinnings of an ordinary language approach.

ally copresent when we focus attention successively upon different things appearing within that field. Such focal attention is central to human psychological functioning. We are passive to what we receive from the environment, but we are active in paying attention in various ways that we will consider throughout this book.

Another aspect of the field of vision as the condition for seeing colored objects is light filling what, in fact, is only apparently empty space between the eyes of the seer and what is seen. A little reflection shows that the apparently empty space between me and the computer on which I am working is full of the sound of the music that plays in the background, though it is not seen. Think also about television and then about the irradiations that form the electromagnetic spectrum. All of that is streaming through the environment within which vision operates, not seen, invisible, and existing precisely as a condition for seeing anything. Paradoxically, could we see all that is "under" or "behind" that phenomenally empty space, we could see nothing because there would be no phenomenal distance between seer and what would be seen. Seeing operates in function of the organic need to identify what fosters and inhibits the ends of the organism. To do that the organism must be built to screen out those features of the surrounding space that would inhibit seeing the things we need to see across that space that gives the illusion of emptiness.

Plato provided a great image for reflecting upon this situation. He said we are like prisoners in a cave, chained so we cannot turn our heads but only stare straight ahead at the cave wall.[3] Behind us is a wall and behind that a fire. And what we see are shadows of things carried on the wall behind us and projected upon the wall in front of us by the fire. Knowing what is known

3. Plato, *Republic* VII.514Aff.

through simple reflection, but even more astonishing through the immense underlying complexity that physics and physiology continue to unveil, and meditating on Plato's image in relation to all that, provides a place of reflective distance for us to consider the oddity of what we take for granted. We live within the way the senses enable us to be tuned in to the things we need in the environment. But the senses' way of revealing conceals what does not lie within their range. Not animals, but only we, as reflective beings, can come to be aware of that and how strange and marvelous the taken-for-granted actually is. The senses as it were blow a luminous bubble, opaque to sensation on the outside, within which we live our unreflective lives. Even as we reflect upon its strangeness, it continues to present the solid sensory world: full sensory actuality with no crack of negativity such as the underlying but invisible powers of the things seen and the in-principle unseeable awareness itself.[4]

Two rather odd things also characterize how things are seen. Objects shrink in size as they recede from the viewer's position, so that each three-dimensional thing is perspectively distorted. A one-foot cube lying on a table looks square on the face frontally and statically viewed, with the top and one side appearing as somewhat flat rhombuses. What is odd is that we do not typically notice the distortion. That is because the psychoneural system that underpins the field of awareness automatically discounts it by reason of having learned from the past what the actual shape and size of the objects are by our having moved around objects or by having handled and rotated them. Our perceptual system has built up within us anticipatory motor schemes that direct our attention away from the fac-

4. This is the realm Sartre calls "Being" that is "full like an egg" and set over against the "Nothingness" of consciousness; Jean-Paul Sartre, *Being and Nothingness*, trans. Hazel Barnes (New York: Philosophical Library, 1956), lxvi.

tual distortions as we move through the environment. That fact of prepersonal learning on the part of our nervous system from the showing of things in the environment opens up a region of experience that must be explained wholly in terms of the nervous system and its interplay with the level of awareness that emerges from it. What the neuropsychological system learns is prepersonal, providing the manifest context within which each of us freely directs her- or himself, let's say, to engage in research into neuropsychological functioning. In doing so, we initiate a programming of ourselves that we freely take up and freely continue. As we rely upon the automatic learning by our neuropsychological system to negotiate our environment, so we also rely upon our previous self-programming to advance in our chosen fields of research and practice.

Besides the spontaneous assimilation of perspectival distortion, the other oddity is that everything appears within a horizon that is the limit of the field of vision. The horizon is like a psychic hoopskirt that goes out ahead of and around us and that moves as we move. The horizon is not *in* me, but is *out there* as the limit of the field of vision. Developing perspectival drawings consists in elongating the lines of things like a cube that stand progressively away from the viewer in the direction of a point on the horizon where the lines would meet. This allows the progressive shrinking of objects to be represented as they appear progressively away from the viewer.

These two considerations, of perspective and horizon, join a third: the question of the status of color. The position that emerged since Galileo was that color, along with other sensory features, are effects within awareness arising out of the brain and caused by wave agitation across the space surrounding one's body—exactly the position naturally suggested by developed knowledge of the physiological reception of light. But as Berkeley pointed out, since we have access to the wave agitation only through the color in our visual field, wave agitation is as subjective—that is, essentially related to perception—as the allegedly merely subjective sensory features.[5]

Of course, the position of common sense is that color is clearly presented "on" things "out there." Well, that means that we are not seeing things in the back of our brains. The only way we could have gotten the data that show us the wave agitation is from their registration on the scopes "out there," presented across visual space on the screens of mechanisms that scientists have constructed to register in our visual field the effects of light waves. Scientists, both in their *practice* looking at and manipulating the objects appropriate to their field and in their everyday lives, are common-sense realists. It is only when they buy into the enclosure of awareness inside the brain that they undercut philosophically the basis for their evidence.

There is a third position, which, I would claim, is actually that of both Plato and Aristotle: that sensory features are a special kind of occurrence *between* perceiver and perceived.[6] Aristotle's was the now familiar question: if a tree falls in the forest and there is no one to hear it, is there sound? His answer was that there is agitation of the air, but no sound until it strikes a hearer.[7] And considering the senses as such, he further noted, the sensory power when it is activated has a distinctive *cognitive identity* with the sensible thing as its sensibility is activated. He said, "The sensory power in act is the sensible thing in act."[8] Like common sense, this position holds that we are cognitively outside our physiological inside, but in a distinctive kind of relation: one of *manifestation* or *appearance* that is sui generis

5. George Berkeley, *Three Dialogues between Hylas and Philonous* (LaSalle, Ill: Open Court, 1945), 38.

6. Plato, *Theaetetus* 153D.

7. Aristotle, *On the Soul* III.426a.

8. Ibid., III.429b27.

and that co-modifies the seer and the seen. That co-modifying act used to be called *immaterial*—but, of course, that depends upon what is meant by "matter."'

But to return to the status of sensory features, the Galilean position that what we see are images inside our brains cancels its own presupposition: we must be cognitively outside our physiological inside, else we could not know what happens outside in examining the physiology of some organism other than our own—and, indeed, even our own as a visual object. The upshot of the analysis is that color, like perspective and horizon, is neither wholly objective nor wholly subjective, neither simply "out there" nor simply "in here." The color of things exists as a peculiar relation between the seeing subject and visible object: that of *manifestation* or *appearing*. And that is the real heart of the nature of seeing: it brings us *cognitively* outside of our physiological inside in the mode of *manifestation* that is sui generis, totally different than what appears as object of seeing or of any other sense. No imaginative picturing can display the peculiarity of the non-seeable character of seeing. We can see the seeable, but we cannot see seeing, though we are intimately acquainted with it.

I want to consider further what makes such manifestation possible. To arrive at that, I want to consider the nature of touch. What will seem like a detour will actually bring us to the crucial insight. Unlike other senses, touch has no localized organ. Rather, the whole organic surface with its subcutaneous sensors is the organ. This requires that the one who can touch and feel itself being touched is cognitively, consciously "in touch with" its own organism as a functioning whole. This "being in touch" occurs when the "consciousness switch" in the reticular activating system of the brain stem is in the "on" position. When that occurs, awareness suffuses the functional body. I say "functional" because being awake is not being in touch with one's brain or one's liver, much less with one's cells, but with what is involved in the self-direction of the organism. One feels awake and thus can feel something that one touches and that touches oneself. Being awake and feeling are equivalent: feeling is nonreflective self-presence.

Relevant to our discussion of seeing, here is the point arrived at through considering touch: what makes the manifestation of something other than oneself possible is the manifestation of oneself as that nonreflective self-presence, other than which the other is manifest. It is self-presence that is the ground for the appearance of something in its being other than that self-presence. Though it may sound gobbledegookey, it is the case: in being focally aware of something other than myself, visually of a set of colored, extended things in the environment, I am nonfocally present to myself as the one who is seeing. In the act of seeing, I know that it is I who see. Being aware at all is a mode of felt self-presence: in animals it is the self-feeling of appetite and of satisfaction or frustration, of pleasure and pain when the animal is absorbed in its relation to the environmentally given. But in seeing and in sensing generally, one's conscious self is not the focus of one's attention; the focus is rather upon the objects given "out there" in the environment. Consciousness is a self-present nonobject that is the locus of the manifestation of any object—not "in" oneself, but "out there" in the environment. We live for the most part in animal extroversion. What makes that possible is the causal series of physical and physiological events whose integral functioning is a necessary but not sufficient condition for seeing. What is sufficient is the felt self-presence of the sensing being.

Focused upon and thinking in terms of visual objects, I said we are "naturally" inclined to think of an organism that sees as confined within the seeable limits of the organism. By reason of tracing the physiological route of transmission of the effect of light upon the retina and then upon the optic nerve, scientists find that seeing occurs in the visual cortex, and, "naturally," what is seen is an image that results from the transmission. Beyond the seeable inside and outside, there is nothing—or so we are naturally inclined to think. Seeing leads us on pragmatically, but systematically misleads us speculatively. We are seduced by the richness available through seeing, instrumentally aided or not, and the marvelous things that such knowledge enables us to do, that we neglect to notice that what is deepest—namely, our own awareness—is not available through seeing it. What is known is not reducible to a set of visual objects or to imaginative models fashioned upon visual objectivity. We are seduced by visual models into neglecting the true character of the act of seeing. Seeing itself is a different kind of "object" than what is seen; in seeing, the being that sees is brought *outside* its visually observable insides, outside its brain, neural networks, and eyeballs. Grounded "here" inside the organism and in the visual cortex, seeing nonetheless brings the seer *outside* its own visually available inside to be cognitively "with" what is seen outside, given precisely *as outside* the seer's body and as *other than* the awareness that sees. Seeing itself is not seeable and thus not imaginable in terms of some visual model.

Let us add another feature. Consider the ophthalmologist who greets her patient. They meet and exchange pleasantries. She then gets to work, focusing upon the eyes of the patient. In looking at the eyeballs, she *looks away* from the looking of the patient and looks at the full visual actuality of his eyeballs. Her work entails *abstracting* from the concrete presence of the patient, no matter how sensorily "concrete" such experience is. In meeting one another, the eyes of the other appear within an overall facial gestalt that is gestural, expressive in character. In living interchange, one does not focus upon the sensory as such but upon the matters about which they are conversing. One enters a space of communication in which sensory features play only a subsidiary role. In fact, focus upon some sensory feature of the other breaks the living connection. In living interchange, the state of mind of the other is expressed in the sensory presentation that is strictly subsidiary to that expression. We see that the other is sad or tired or cheery or just in an ordinary, everyday mood. The eyeball as a visual object is subsidiary or nonfocally present within one's attention to the person as a live subject, present primarily through the expressive appearance of the face. According to an old Platonic saying, "The eyes are the windows of the soul."[9]

There is a fundamental distinction then between the *lived body* and the *body-object*, the body as felt and lived "from within" by the person who is embodied and that same body looked upon by the scientist. But there is also a distinction between the body objectified in scientific investigation and in medical practice and the body as *expressive* of the disposition of the person whose eyes are being observed scientifically. Expressivity indicates to another the lived body. Physical science systematically abstracts or looks away from expressivity and thus from *meaning* or how a given thing is to be *read* or *interpreted*. One could consider any living body as *aimed* at its full flourishing and at the endurance of its species.

Mechanistic science systematically looks away from teleology or goal-directedness as a fundamental characteristic of life, forgetting that mechanisms themselves are devised for reaching goals. And so we get a fundamental distinction between fact and value: facts that we can see and values that *we* are said to attribute to things. However, what is given in the case of living things is not simply factually observable mechanisms but also goal-directedness on account of which organisms can succeed or fail. So, when we find living things, we find values: on the one

9. Plato, *Phaedrus* 255C.

hand, there are the *intrinsic values* of growth to maturity, sustenance, and, following the attainment of maturity, reproduction and nurturing of offspring. On the other hand, there are the *extrinsic values* of things in the environment that serve the intrinsic values, as a given organism may also have extrinsic value for its mate, its parent, or its prey. So human goal projection is not a stranger in a strange land, but a carrying on at a level of deliberateness what had already been going on in both phylogeny and ontogeny, in both the coming into being of the species and the development of the individual human organism.

Parallel to the abstract looking away from meaning is what is involved in the act of reading the E-chart. When the patient scans the E-chart, the letters have been taken out of their typical function of providing a text for reading and are put together in such a way as to invite being focused upon all by themselves, as in the case of the eyes for the ophthalmologist. In reading an actual text the letters are, like the eyeballs of other persons in our encountering them, strictly subsidiary to the activity of following what they express: the meaning indicated by the conventional configurations. Focusing upon the eyeballs and upon letters defocuses the expressive meaning of the look of the other through the eyes and the meaning of the letters in words. We abstractly attend away from the meaningful interchange within which we live our lives.

Further, in the ophthalmologist-patient relation, while looking at the patient's eyes as visual objects, the doctor continues *talking* to the patient. In examining a patient—for example, for lens prescriptions—the patient's verbal input is essential. He is asked which in each of a series of two lenses yields a clearer vision of the E-chart.[10] Here we have sound available through

10. One of my colleagues wondered how an ophthalmologist might correct a dog's vision.

one's ears functioning subsidiarily in carrying the meaning of the words, like the visible eyeball of the other is subsidiary in relation to one's focus upon the gaze of the other or like the printed letter subsidiary to reading. The primary presence of the person, going along with his or her gestural style, is in what he or she says. This expresses not primarily the *disposition*, but rather the *mind* of the patient, what he or she thinks and out of which she or he lives. So the human body that we see is not simply a visible object upon which we can act, guided by science; the human body is first and foremost the expression of the dispositions and thoughts of the person, or, in Hegel's terms, the human body comes to be penetrated by, and expressive of, Spirit.[11]

Further still, the doctor is not just looking at the full actuality of the eyes but focusing on them in function of an attempt to understand what is present as a symptom of what she intends to fix. She knows what constitutes regularity of structure and function and can take notice of deviations from typicality. She not only sees the eyes but sees them *as instances of types*. She is focusing *with developed intelligence* that recognizes patterns that call for standard procedures in bringing the eyes back to normality of functioning. She is a skilled practitioner of ophthalmology. She brings her intellectual experience as a practitioner to bear upon her sensory experience.

In the development of her skills, she uses optical instruments: scopes that magnify the visual objects, devices that test peripheral vision, glaucoma testing devices, and the like. All this is possible because those who first invented such instruments knew the laws of the relation between light, diffracting media, and visual presentation. This entails advancement beyond the skillful coping of animals with the visual environment

11. G. W. F. Hegel, *Hegel's Philosophy of Mind*, trans. Michael Inwood (Oxford: Oxford University Press, 2007), §411:147.

driven by organic need. It entails advancement beyond the immediacy of visual givenness of *individuals* in the environment to the *universal types* of structure and function that are the objects of scientific investigation, independent of organic needs. Though the physician is interested in curing the ailments of her patients, understanding types of structure and function can occur apart from that interest and in function of the pure desire to understand and even stand in awe of the miracle of seeing, its astonishing difference from the things we see.

It is important but a bit more difficult to notice that there are several features of sensory objects in general, and not just visual objects, that help us to understand better the character of seeing. Sensory objects are all *individual* and *actual*. That means that what we cannot visually see is the *potentiality* or the *power behind* or *beneath* the manifest surface. And what is peculiar about the native power of any entity is that each individual power—for example, my own power of seeing—is not simply individual but is a *universal* orientation that requires individuals of a natural *kind* of object in the environment. Seeing as a power is oriented toward *all* individual features of the kind we call "colored." And the capacity to be seen is oriented toward *all* organisms that can see. Now, kinds are universal: a kind is found in *all* individual instances. But sensory objects are all individual. We have the capacity we call "intellect" to move beyond individuals and recognize the universality of the orientation of powers and the universality of object-types as such. That is why we can have science and, indeed, why we can have language where the sound visual shape of a word is a sensorily available stand-in for the intellectual apprehension of the type in a concept. Concepts function like glasses: we normally and unreflectively look *through* them and not *at* them. But we are now attending reflectively *to* them.

What is true of native powers is true also of acquired skills. A skill is a capacity to deal with *all* the individuals of the kind correlative to the skill. If I become a skilled carpenter, I can deal with all individual instances requiring the work of a carpenter correlative with the level of skill I acquire. An ophthalmologist is one who has mastered the skill of curing problems with the eyes *as such*—that is, not just the eyes of the patient with whom he/she is dealing in any given instance, but of any and every individual who comes to the doctor with eye problems. The universality involved on the part of the skilled physician and the patients whom she cures is not something one can see, but clearly something one can and does come to know—even though one might say, metaphorically speaking, "I *see* what the problem is."

In summary of some of the major features already noted, what cannot be visually inspected or imaged is seeing itself and the overall field of awareness involved, the powers that underlie or stand behind visible actuality, and the universal types we can come to recognize within each sensory field. What cannot be seen is the mind of human beings, which is nonetheless present to us in the speech and action that we can hear and see, and which is present as the mind of the one who sees, hears, thinks, acts, and speaks as a responsibly self-present human agent.

These observations give us new ears to hear the claim of Aristotle: if the eyeball were an organism rather than an organ within an organism, its soul would be the power of seeing, and its act would be factual seeing.[12] The holistic character of awareness expresses the holistic principle that underpins it as a metabolic process furnishing the instruments for seeing. The power of vision is not a feature of a kind of ghostly substance haunting the nervous system; it is a second-stage emergent out of a prior and enduring metabolic process. The ground of that process and of emergent consciousness is what Aristotle called the "psy-

12. Aristotle, *On the Soul* 412b 19.

che" or "soul." Francis Crick is looking in the wrong direction to find the soul, which he considers a hangover from antiquated ways of thinking.[13] Crick's thinking is rather a degeneration of a richer tradition that he chooses to ignore. The soul is found operative in the power of awareness underpinned by the constructive power of the organism.

Let me add one more observation—a bit more difficult to see, but of primary importance in human life. My first consideration is easy to observe, but it leads to the more difficult point. Beyond visual functioning and coping with the environment, science, through seeing and manipulating, can aim at understanding the nature of organism as such and at a general theory of how the current inhabitants of various ecosystems have come on the scene. It can aim at understanding the system of the underlying subatomic particles embedded in the space-time-energy matrix that underpins and encompasses whatever appears within the sphere of ordinary awareness. It can aim at a general cosmology that extends to the evolution of the entire space-time system that encompasses everything material. It can even raise the question of the possibility of something beyond such a system. And that very questioning indicates the "being beyond" of the questioner. As Hegel remarked, for any putative limit we can always ask what lies beyond the limit, indicating the all-encompassing character of the human spirit.[14] That is the observation.

Here is the point: science can aim at all that because the human mind is by its nature aimed at absolutely everything by reason of the operation of the *notion of Being* and, along with it, the unrestricted character of the principles of noncontradiction

13. One of the best philosophical works on medicine is Leon Kass, *Toward a More Natural Science* (New York: Free Press, 2008).

14. Hegel, *Philosophy of Mind*, §386, *Zusatz*.

and identity. We know ahead of time that whatever is cannot have and not have a given attribute at the same time and in the same respect.[15] For this to be the case, things and their properties must have distinct identities that endure through time and in place. As a long line of philosophers going back to ancient Greece have noted, what makes the human mind to be such is the *notion of Being.* "Being" includes absolutely everything, for outside being there is nothing.[16] But for us it includes everything only *in principle*, since we can actually include *in fact* only what we are able to discover by using the proper methods and dispositions or to learn from those who did discover things by reading their reports.

At some undetermined point in the development of the human individual, the *notion of Being* arises in the mind and grounds the question about our place within the whole scheme of things. We exist as *the question of Being* as such and as a whole.[17] Such a question is built into our nature. It is that upon which is based our ability to think of meanings that apply always and everywhere in time and space where individuals of a given type may be found. Space and time as encompassing forms that make possible the "always and everywhere" as the field of application for our concepts are themselves a contraction of the notion of Being to the domain in which we are intellectually embodied.

It is the *notion of Being* that gives us inward distance from the here and now where our senses operate and within which we direct ourselves. The *notion of Being* gives us distance from

15. Aristotle, *Metaphysics* IV.1005b.19–20.

16. One of the clearest treatments of the notion of Being is Bernard Lonergan's, in *Insight: An Essay on Human Understanding* (London: Longmans, Green, 1964), 348–74.

17. Martin Heidegger, *Being and Time*, trans. Joan Stambaugh (Albany: SUNY Press, 1996), 12.

what continues to determine us as genetic endowment, cultural upbringing, and the history of our own choices. The resultant of these three factors both opens up and limits my concrete possibilities. But as a responsible agent I am at a distance from all that and must, by nature, choose what to do with myself. Reference to the Whole via the initially empty *notion of Being* forces each of us to take responsibility for what we become.

Further, as Aristotle pointed out, intellect as the ability to apprehend ultimate principles is also the ability to grasp "the ultimate particulars" required for the function of *phronesis* or the capacity for recognizing what is to be chosen in particular situations.[18] The notion of Being includes concrete beings as well as the recognition of universals, or, as Lonergan puts it, the notion of Being includes everything and everything about everything.[19]

I would go further still, for *phronesis* only deals with things insofar as they enter into the situation of human action. But, as Derrida notes, their full reality recedes into indeterminate depth, and full presence is deferred. Science involves an entry into the "more" in things than what appears on the surface of everyday functioning.[20] And beyond that, there is what Heidegger, Marcel, and Buber noted: each thing is a mystery of which we can gain a real sense, beyond everyday functioning.[21] And Plotinus testifies that, for one who has experienced oneness with the One, the phenomenal world appears as the Face of the

18. Aristotle, *Nicomachean Ethics* VI.114bb.1–5.

19. Lonergan, *Insight*, 356.

20. Jacques Derrida, "Différance," in *Margins of Philosophy*, trans. Alan Bass (Chicago: University of Chicago Press, 1986), 7–8.

21. Heidegger, "Memorial Address," in *Discourse on Thinking*, trans. J. Anderson and E. Freund (New York: Harper and Row, 1966), 54–56; Marcel, "On the Ontological Mystery," in *The Philosophy of Existentialism*, trans. Manya Harari (New York: Citadel Press. 1961), 9–46; Martin Buber, *I and Thou*, trans. Walter Kaufmann (New York: Scribners, 1972).

Beloved.[22] Indeed, for one who most fully loves, the beloved contains an ineffable mystery.

Both *intellect*, as the ability to apprehend universal meanings, and *will*, as the capacity to choose, are grounded in the operation of the *notion of Being*. It functions not only to ground our distinctively human operations, it functions to draw us beyond our current hold on things and our current modes of operation to modes of thinking and acting that correspond more and more to our place in the scheme of things.

It is all of this that any specialist as a human being should understand. He or she will thereby be enabled to resist the reductionist temptation into which Crick, like so many others, falls. He or she will become increasingly aware of the way we humans are responsibly positioned within the Whole as scientists, as physicians, as friends, as members of a community, and ultimately as members of the community of all human beings.

IN THIS CONCLUDING SECTION, I want to open up a more wide-ranging battlefront in coming to terms with those who would reduce awareness to material processes. Along the same lines as Crick's, there is an earlier book by Dean Wooldridge called *Mechanical Man* that is a refreshingly frank statement of the reductionist program and its central difficulty. The author deliberately hearkens back to de La Mettrie's *L'homme machine* published in 1748.[23] For Wooldridge, in that work de La Mettrie had issued a promissory note, and subsequent research had suc-

22. Plotinus, *Enneads*, trans. Steven MacKenna (London: Penguin, 1991), II.9.16.

23. We should note that Julian Offrey de La Mettrie was not a pure mechanist; another work, also published in 1748, clearly indicated such: de La Mettrie, *Man the Machine and Man the Plant*, trans. Richard Watson and Maya Rybalka (Indianapolis: Hackett, 1992).

ceeded by 1968 in paying out about 80 percent of the promise.[24] But then, with admirable honesty, Wooldridge admitted that consciousness isn't anything like the focal mechanisms. From what is now known about human mechanisms, all it can be is a purely passive byproduct of the nervous system, "a sort of window through which we can observe a part of the workings of the brain."[25] This says nothing about who "we" are or what "watching" is or that only workings of the brain are observable. John Searle, in his *Recovery of Consciousness*, remarked that, if consciousness has no function, it is a very expensive evolutionary product that contributes nothing to survival.[26] Wooldridge is a typical epi-phenomenalist whose position on consciousness as a secondary phenomenon is a scientific embarrassment. Our reflections upon all the essential features of human consciousness involved in seeing show the way to overcome the embarrassment.

It is the pre-personal functioning of the perceptual system that calls for a neurophysiological explanation. An experiment with prisms that inverted the images on the retina led initially to seeing everything upside down. But after a certain period of time, *the psychoneural system*, correlating the visual images with the other sensory reports developed in terms of the motor habits formed for negotiating the environment, learned to reinvert the images so that they appeared normally.[27] If a photoelec-

24. The figure of 80 percent is not meant to be an exact figure, but indicates that research is nearing the end. If very much was secured at that time, research to the present has advanced significantly further and thus further encourages the reductionist position.

25. Dean Wooldridge, *Mechanical Man: The Physical Basis of Intelligent Life* (New York: McGraw-Hill, 1968), 85, 160–61, 167. Even though this book is nearing fifty years old, the clarity and honesty of his admission remain relevant.

26. John Searle, *Rationality in Action* (Cambridge, Mass.: MIT Press, 2001), 296.

27. The classic locus is George M. Stratton, "Some Preliminary Experiments

tric cell were linked to the proper motor system, such learning could, in principle, be explained without the awareness, although I think that animal/human learning at this level would involve the synthesis of *conscious experience* derived from holistic interaction with the environment. What then is the function of awareness at that level? I think animal awareness must be considered en route to human awareness as having a kind of self-determination, but one governed wholly by biological needs. Animal awareness allows for a more flexible adaptability than is the case further down the evolutionary line, for example, in the case of insects where the character of awareness, if existent, is dim. But distinctively *human* awareness can detach itself from being wholly governed by those needs, though human free functioning also has its neurological correlates. Athletes, ballet dancers, and musicians, through practice, learn to align themselves with the gift of spontaneous performance that they let their bodies execute. The distance of objective awareness and deliberate control (so-called acts of intellect and will) have to be given up in the performance situation so that, as the Zen masters say, *it* acts in someone when he or she achieves "no mind."[28] This alignment with the gift occurs in such a way that the agent simultaneously takes responsibility for what she or he allows to happen spontaneously through him- or herself as a psychophysical whole. Future brain research has to target the neurological basis for that state of maximum free performance where responsibility and necessity, choice and gift intersect.

Reaching the skill level involved in excellent bodily performance involves having freely programmed oneself by practice.

---

on Vision without Inversion of the Retinal Image," *Psychological Review* 4 (1897): 182–87.

28. Daisetz T. Zuzuki, "The Zen Doctrine of No-Mind," in *Zen Buddhism: Selected Writings of D. T. Zuzuki*, ed. William Barrett (Garden City, N.Y.: Doubleday, 1956), 157–226.

There are actually three basic levels of programming upon which we rely in our free self-dedication. First is the genetic level, which delimits each of our particular possibilities. It is open not only to our psychoneural system's automatic learning, which allows us to negotiate the environment; it is also open to our being programmed by primary caregivers, most especially in language use through which all distinctively human learning takes place. The ongoing impact of our interplay with others and our being saturated by communications media continues the programming process. But at some point each of us began to take responsibility for our own programs, choosing to follow paths to various skills. We then became free self-programmers within the parameters of our previous trileveled programming. The future of brain research should show how these programs are formed, stored, and made available to us as conscious agents, always with the awareness that there is a level of programs that we freely take up, many of which, like the pursuit of neuropsychology, we choose to form and will continue to choose.

As a concluding remark, I want to extend our considerations to the widest context of the cosmos. The rational freedom to choose the pursuit of the acquisition of different skills has to find its place in an evolutionary universe. I hold that if we accept "whole-hog" evolution, a view that everything, ourselves included, is a product of nothing but the evolutionary process, what has to be changed is not simply our view of mind as a separate thing, but, in a sense more important but too often overlooked, what has to be changed is our view of matter as other than mind. If we attend carefully to awareness, we see transparently that science is one of the glorious achievements of the human spirit, using one's freedom to focus one's powers in the pursuit of truth. If that comes out of the initial stages of the universe we inhabit, the early stages have to be understood, not only in terms of visual evidences, but also in terms of the unseeable potentiality for the awareness and thus the science that emerges out of it. That places the shoe on the other foot of the reductionists. It is not those who hold for the distinctive character of mind who have to yield the field; it is the reductionists who ignore the peculiarities of mind and who are popularly thought to command the field.

The view of mind changes the view of matter as well as of mind as a separate thing, but it does not change the immediately given view of mind as a center of self-directive intelligence oriented toward the Whole. The king and queen of philosophic reductionists, Paul and Patricia Churchland, whom Francis Crick especially admires, have no clothes.[29] Like Crick, they fail to attend to their own practice, aimed at the Whole and freely choosing a reductionist program. The big fuss about reductionism that has unnecessarily dogged science and that appears all too often in the popular press is based upon abstracting from the practice of scientists and the intelligent and responsible activity of ordinary human beings. The reductionists are involved in a performative contradiction, using their rational freedom to argue against such freedom.[30]

So, reflection upon what is present in awareness when we see gives us a basis for coming to terms with the all-too-common tendency to reductionism that follows in the train of scientific developments in neurophysiology. Our presentation has focused most basically upon the free self-determination that follows from our being referred, beyond our current state, to the encompassing Totality. Such reference founds the capacity for apprehending the

29. Paul Churchland and Patricia Smith Churchland, *On the Contrary* (Cambridge, Mass.: MIT Press, 1998).

30. This lays the basis for the cosmology I will sketch out in chapter 4. See also Robert E. Wood, "Five Bodies and a Sixth: On the Place of Awareness in an Evolutionary Universe," in *The Beautiful, the True, and the Good: Studies in the History of Thought* (Washington, D.C.: The Catholic University of America Press: 2015), 312–23.

universal that is operative in ordinary language and in the specialized languages of the sciences; it founds as well our capacity for freely committing ourselves to searching out our place in the Whole, and especially the role of our neural systems in underpinning that search.

2 First Things First

On the Priority of the Notion of Being

IN THIS CHAPTER I want to consider three propositions:

1. "First to arise within intellectual awareness is the notion of Being."[1]
2. The human being is defined as "rational animal."[2]
3. Knowing involves "the complete return of the subject into itself."[3]

What are the conditions for the possibility of recognizing what is involved in those claims? In exploring them, I want to open up a way into Parmenides, Plato, and Aristotle, as well as Aquinas, Spinoza, Nietzsche, Hegel, Heidegger, and Derrida—a rather ambitious task.

1. "Primum quod cadit in intellectu est ens"; Thomas Aquinas, *Summa theologiae* (hereafter *ST*) (St. Louis, Mo.: Aquinas Institute, 2012), I, q. 5, a. 2.

2. Aristotle, *Politics* I.2.1258a10.

3. Aquinas, *On the Truth of the Catholic Faith: Summa contra Gentiles*, 4 vols. (New York: Doubleday, 1955), IV, 11 (henceforth *ScG*); see also *ST* I, q. 84, a. 6.

BUT INSTEAD OF weighing in at the deep level, let us begin very simply and immediately with what is functionally, though not theoretically, indubitable but that is also necessary for communicating the approach to and articulation of the theoretically indubitable. Let us begin with what is first—namely, the word "First," and with what is first in that—namely, the capital letter *F* as it appears on the printed page. After all, we could never practice Cartesian doubt, or at least know that we are doing so, without Descartes's having written about it. Even more, Descartes could not have formulated the method and the successive elimination of the dubitable without the French or Latin he employed in the application of the method and thus without knowing his *ABC*s.

Even though the three propositions listed are grand in their scope, I prefer that we begin now at the most obvious and most humble level that in our grander aspirations we are prone to ignore. I begin with it because it is always already given but usually overlooked when we carry on our literate lives.

The capital letter *F* is an immediately given visual object, given outside in the environment where we each, alone and together, conduct our lives: it appears on a white page separate from its viewer across phenomenally empty space filled with light. It is a visual object on its empirical outside, but on its intelligible "inside" it carries a universal function as an alphabetical unit whose job is to enter into the construction of words (or function in the E-chart for visual testing). *F* is part of an alphabet, a constructed eidetic system, a system of conventionally fashioned universals whose meaning is invariant across all empirical instances, no matter how different they might be empirically. It appears visually on the current page as the uppercase of Minion *F* and subsequently in lowercase *f* several times. It could also appear in Meta lowercase f and uppercase F or in any other typeface or handwritten script, empirical variations of which are endless.

What is remarkable, if we give it some thought—banal if we do not—is that the *f* is not only *identical* in all instances of its visual presentation, it is *empirically different* in each case of its being written, obviously if the typeface is different, but even when it is the same. Both ingredient in and apart from its different modes of empirical instantiation, its function is recognizable as identically the same. That seems to give the lie to nominalism at the very level of its names: we apprehend constructed universals that transcend their empirically available instances as a type of apprehension that operates in linguistic use generally and that we have come to call "intellectual."

The conditions for the possibility of such operation is our apprehension of the universals involved in the universal orientation of our own *powers* and their corresponding *kinds* of objects and the *kinds* of things that exhibit them. Each power, such as vision, is such by being oriented ahead of time to *all* instances of the *kind* of feature correlative to the power; in the case of vision, it is color. Seeing, though individual in each seer, is *universally* oriented; but though universally oriented, it is actualized only in revealing *individual* instances of its corresponding type of object.[4] This undercuts, at the level of the existing individual, the alleged split between the universal and the individual, the healing of which can only be recognized through the kind of power through which the universal is recognized as such: the power of intellect. Note also that a biological reproductive line is the carrying on of a *kind* that we can recognize as such and thus as the same in essence in all the differences in its empirical instances. These last two observations—regarding powers and reproductive lines—pose the most fundamental challenge to nominalism.

4. G. W. F. Hegel, *Hegel's Philosophy of Mind*, trans. Michael Inwood (Oxford: Oxford University Press, 2007), §401, *Zusatz*.

Return to our starting point with the letter *f*. It is part of an alphabetical system based upon a phonological analysis of spoken words. The original analysis identified five ideal instances of the ways in which the human oral cavity emits sound in speaking a given language: the vowels *a*, *e*, *i*, *o*, *u*; and the nineteen ways in which these sounds can be clipped through what "sounds together with" the vowels, namely, the consonants (from the Latin *cum* and *sonare*): for instance, sibilants or "hissings" *c*, *s*, *x* (at least in the sound identifying them as separate alphabetical units) or labials pronounced by the lips (Latin *labia*) *b*, *m*, *p*, and the like. It took an act of genius to perform the phonological analysis that covers all possible variants in a given language. It is a genius parallel to the one that discovered that all known and future metrical regularities could be deduced from a few axioms and postulates in the Euclidean system; or the genius that discovered the Periodic Table of Elements.

The first philosophical experience reported by Paul Weiss, the dean of twentieth-century American metaphysicians, was when at age six he was introduced to the alphabet. He was astonished at the fact that everything that can be said can be said by using just twenty-six signs. Of course, there are different modulations in the phonological value of the letters or sounds depending upon the words in which they appear. For example, how *-ough* is pronounced depends upon the many different words in which it appears: *through* (ou), *rough* (uf), *though* (o), *cough* (awf), *bough* (ow); but, surprisingly, there is no sound for what that combination of letters immediately suggests: *-ough* (oog). So the alphabetic units have a somewhat loose relation to the sounds they were invented to indicate.[5]

5. Given the irregularity between the written and spoken word, there could be two ways to spell *fish*: "ef, eye, es, aych" and "*ghoti*": "gh" as in *laugh*, "o" as in *women*, and "ti" as in *composition*.

Vowels and consonants combine into words that have a set of possible meanings determinable by definitions that fit a given word into a whole network of meanings. We are able to unpack the words we use in each definition in an ever-expanding network that extends to the whole of that about which we can speak.

Focus then upon the meaning of "First" as the first word in the first proposition. It could be defined as the beginning of a numerical series, where "numerical" refers to a series of identical units determined by regular increments and "series" is described as a type of sequential order, "order" as a coherent arrangement of factors, and "type" as a universal over against its instances; "coherent" in turn defined as …, then "arrangement" defined as …, "factor" as …, "universal" as …, "of" as …, "in" as …, "a" as …, "where" as …, "could" as …, "defining" as …, "or" as …, "network" as …, "refer" as …, "be" as …, "as" as … *et sic ad systema verborum completa*. The definitional network eventually is seen to cover the whole linguistic system, which itself refers to how the Whole within which we find ourselves presents itself within a given linguistic tradition. We could go back to the original definition of "first" and follow out not only the word "numerical," but also "definition," "beginning," and all the other accompanying words. But of course, as the definitional net widens, there could be different ways of defining the terms: here is where hermeneutics and hypotheses enter the picture. Each definition involves its definers, which in turn have to be defined, etc. The attempt to work out the place of any given thing in a definitional system is a helpful exercise in making explicit the implicit understanding involved in language use.

The situation is complicated by the fact that many words have more than one meaning—the OED lists a massive set of meanings for the word "heart": eight pages, twenty-three columns. (That, I would maintain, is because this term lies at the heart of language; and that is because its referent lies at the heart of our lives.)[6] Which meaning is being employed can only be determined as it enters into a given sentence. A sentence as a combination of words pins down a possible plurality of meanings to the function they take on in specific propositions.

What makes a proposition out of a string of words is, fundamentally, the combination of nouns and verbs that anchor, inter alia, adjectives and adverbs. The combination makes a truth-claim in the three declarative sentences we are examining. And the truth-claim of a proposition is verified insofar as it is revealed in judgments as corresponding to what is the case. In the current chapter I claim that the propositions I have generated thus far correspond to what is the case with regard to written language.

Of course, the phonological analysis indicates that what writing presupposes is *speech* in which the function of written units can be identically instantiated in sound. The text is something to be *read*, either out loud in my delivering it to an audience or in an individual's silent subvocal reading. So the alphabet and its employment in sentential functions bridge two fundamentally different sensory media: visual and audile. Writing translates the temporal flow of speech in sound into the spatial fixity of a seeable text. When one reads what is written or hears another read it, there is a translation from the spatial arrangements we call "writing" to the temporal unfolding in reading, reversing the original relation where the temporality of speech was translated into the spatiality of writing, replicating *ad extra* what originally occurred as the transcription of the temporality of sound into the spatiality of the brain.

The identically same meaning now appears through sound

6. See Robert E. Wood, "The Heart: Phenomenological and Historical Studies" (in manuscript).

generated from the mouth of a reader across the space within the range of audibility for its audience—or it is subvocally pronounced by the silent reader. It appeared in the audio mode when the writer addressed its original audience. It can be sounded out by voices with different pitches, modulations, and timbres. The wonder here—and the basic difficulty for a thoroughgoing empiricist—are that the *identically same* meaning is found, not simply in variation of kinds of print and their instances, but in the *completely* different medium of sound produced by empirically different voices. We spontaneously recognize identical sameness in differing empirical instances of differing kinds within a given medium and within a completely different medium through all the empirical modulations in each medium.

We might add here a consideration of Braille where the spatial fixity involved also involves another radical difference in the medium: textured surface accessible to touch; yet the same functions are identical across the three media and the differentiated appearance (minimally spatiotemporal location) within the same medium. We might further add the visual communication of meanings, apart from writing, in sign language. Here is the same visual medium as statically inscribed writing, but in the moving temporal form of hand signs.

Further, what speech and writing presuppose is induction into a specific linguistic tradition. Sentences in a given language admit of indeterminately many instantiations, but they also admit of being translated—*mutatis mutandis*—in indeterminately many different languages, such as our initial sentence, an English translation from a proposition in Latin of Thomas Aquinas: *Primum quod cadit in intellectu est ens*—"First to occur within intellectual awareness is the notion of Being."[7]

7. Translation mine.

It is with poetry especially in mind that I added the *mutatis mutandis* qualification for translation. In each culture, poetry, as the language of indwelling in the life-world, draws upon connotative—that is, associative—rather than simply definitional meanings of words as it draws upon the sonority of its language to provide another kind of meaning: emotively associative meaning. But there is also variation in the conceptual network defining each term employed both in differing cultures and within a given culture, complicating both translation and more complex communication in any language.

So propositions or the translatable truth-claims transcend instantiation in ethnically specific sentences. In principle, the cognitive meanings involved are accessible to any human being, across all linguistic differences. The translatability of language indicates our identity in nature with all other humans who individually and culturally have their own identities. We are all identically human in nature but not empirically identical human beings.

Continuing now with the Latin, note that the Latin term *cadit* in *Primum quod cadit* introduces us to another kind of meaning: that of metaphor. The first meaning of *cadit* is "to fall"; but here, as metaphor, it refers to an occurrence *within* intellectual awareness. It is important, especially given the original meaning of *cado*, that the *in* is properly translated to correspond to Aquinas's intention—but, more important, to what is the case. *Cadit* suggests coming from without; but in Latin, the *in* with the dative *intellectu* indicates rather what happens *within* intellectual awareness. The notion of Being arises a priori *within* intellectual awareness and does not come from without. That is a hermeneutical point about translating Aquinas. We will later consider the truth and implications of the content of the proposition.

Continuing with the metaphor, the word "metaphor" is also a metaphor whose original meaning is "carrying over" (from the Greek *meta* and *phorein*). *Intellectus* and its derivative English "intellect" are metaphoric as well, though here, contrary to the case with *cadit*, the metaphor is effaced and is not necessarily known in our following the meaning. According to Aquinas's etymological guess, *intellectus* comes from *intus* or "within" and *legere* or "to read": intellect is the capacity through which one reads universal meanings within the sensory.

Such meanings are, first of all, the universal meanings of immediate *sensa*: consider the whiteness of the page and the blackness of the words written on it or the peculiarities of the sound of the voice as one might read the written words to an audience. Universal meaning is also expressed and therefore has to be interpreted through gathering the meaning of the words through the sensa, through the written page. We will return to the notion of intellect later in terms of the truth of what is being said in the proposition about it that we are considering.

Looking back over what we have distinguished, we find a hierarchy: the *alphabet*, the empirical instances of *sights* in writing and reading and sign language, of *sounds* in speaking and hearing, in *textures* of writing in Braille; then, after the alphabet, the *word* composed of alphabetical units; then the *sentence* composed of words; then the *proposition* distinct from ethnically specific sentences, though only accessible by being imbedded in a particular language; and finally, the termination of the series in the *judgment* of truth or falsity.[8]

WE BEGAN TOGETHER outside each of our own inwardnesses fixed "outside" on the words on a page or on the sound of a voice reading them. In so doing, as we automatically do, we drew upon

8. See Bernard Lonergan, *Insight: A Study of Human Understanding* (London: Longmans, Green, 1957), 271ff.

our own native English. In doing so we are situated outside of our personal privacy in a space of common meanings as language incarnates our inner apprehension of meaning in sensory media, in the sights (written or signed) and the sounds or embossed textures of our language mediated by and centrally mediating our particular cultural life-world. Language as sensorily embedded bears witness both to the essentially incarnate nature of our thought and to our inhabitance of a public space of meaning. We live our lives outside our conscious inside in the public space of meaning presented through encounter "out there" in the sensory surround.

Meaning in this sense involves the apprehension of *universality* carried by the sights and sounds and textures of language. To pursue further the notion itself, note that it involves being beyond the here and now of bodily encounters, open to space and time as encompassing wholes, for we always implicitly understand that the letter *f*, in our opening example, as well as the words and sentences into which it enters, can each be instantiated *any time* and *any place* where its instances can be produced and thus encountered.

Space and time are encompassing forms present in all human wakeful life, empty without empirical filling and beyond any filling we achieve. Infamously, Bishop Usher in the seventeenth century, working in terms of biblical chronology, went back to 4004 B.C. as the reputed beginning of time. Now, through patient empirical observation, inference, and instrumental construction, our awareness of time has expanded to some 13.75 billion years. And, in parallel fashion, our awareness of space has expanded empirically from the azure dome above—who knows, maybe the staggering equivalent of a month's journey!—to some 92 billion light-years across by current estimates. The forms of space and time, since they always refer us to what lies beyond and encompasses empirically filled space and time, cannot derive from experiences of particular spaces and times but, as Kant rightly noted, are brought to particular experiences as a priori to any conceivable humanly conscious operations.[9]

This points to the ground of our considerations thus far in the second a priori that appears in our first proposition: the notion of Being. Space and time, as forms we bring to bear upon experience, specify ahead of time the context within which human intellectual awareness operates. The reference to the whole of space and time is grounded in a reference, provided by the notion of Being, beyond empty space and time to include all instances found within those forms and to everything about all instances. Though it is only actualized when throwing light on what is sensorily given from without, the notion of Being cannot come from without, because it refers us to the Whole of what is, for outside Being there is nothing.[10] That is why for the *primum cognitum*, for what first occurs in intellectual awareness, Aquinas used the phrase *in intellectu* rather than *in intellectum*, the latter indicating coming *into* the intellect from without, the former indicating an arising *within* intellectual awareness.

The notion of Being, making its appearance in human awareness, has absolutely unrestricted universality: it covers everything that is and everything about everything. Its unrestrictedness is indicated by our ability to ask, beyond any putative limit, what might lie beyond that limit.[11] It even opens up the question whether there might be something that exists beyond space and time in absolute difference from anything finite, something we

9. Immanuel Kant, *Critique of Pure Reason*, trans. Norman Kemp Smith (New York: St. Martin's, 1969), A20/B34 to B73, 65–91.

10. For an insightful presentation of the notion of Being, see Lonergan, *Insight*, 348–74. For an analysis of the way in which that notion grounds every distinctive operative property of humanness, see Wood, "Being Human and the Question of Being: On the Unitary Ground of Individual and Cultural Pluralism," *Modern Schoolman* (Winter 2012): 53–66.

11. Hegel, *Philosophy of Mind*, §386 *Zusatz*.

could never imagine—but then we cannot even properly imagine our own awareness.

The notion of Being is in cognitive identity with thought:[12] to think is to think Being, beyond all appearances and beyond all finitude—beyond all appearances, because it refers to the *whole* of each thing only partially revealed through our sensory filters and cognitive inferences. Animals live wholly in relation to the sensorily manifest surface guided by their organically based appetites. The notion of Being also directs us beyond all finitude, because we can ask of any putative limit what lies beyond the limit. And the first answer is that, as intellectual beings, *we* lie beyond the limit. Whether there is something else beyond finitude is another and quite significant question—the most significant of all.[13]

As we initially awakened, at whatever age, to intellectual apprehension, the notion of Being, like the notions of space and time that it grounds, was empty. It was, as Nietzsche noted and Hegel had already accepted, "the last trailing cloud of evaporating reality."[14] It expresses the initial emptiness of our orientation toward the Whole. However, the notion of Being actually has additional content, a formal one that guides all rational activity. Along with the notion of Being there arises within intellectual awareness the principle of noncontradiction that has the unrestricted universality of the notion of Being that it articulates. As Aristotle formulated it, "A being cannot both have and not have the same property at the same time and in the same

12. The first to articulate this notion is Parmenides.

13. Sartre considered human consciousness to be a lack "beyond Being," simply other than bodies as beings and not related to anything else beyond; Jean-Paul Sartre, *Being and Nothingness*, trans. Hazel Barnes (New York: Philosophical Library, 1956), lxv–lxvii.

14. Friedrich Nietzsche, *Twilight of the Idols and The Anti-Christ*, trans. R. J. Hollingdale (Baltimore: Penguin, 1968), 37; Hegel, *The Science of Logic*, trans. A.V. Miller (New York: Humanities Press, 1969), 82ff.

respect" (*To gar auto hama huparchein te kai me huparchein adunatov to auto kai kata to auto*).[15]

Though it is always operative, we come to awareness of it only through developed experience. In order to formulate the principle, we need the sensory appearance of things and the operation of language we have already analyzed, its flow through speech, and its fixity in meaning. But several other factors are involved. Recognizing "a being" implies its being *one* of *many*. Further, we identify a being by its endurance over time in displaying the sensory features as immediately given properties that our sensory system gathers together in *phantasms*, modes of holistic and multi-faceted appearance. "Can" plays in relation to "is" as *possibility*; and "not" involves a negation that here becomes *impossibility*. "Property" involves an aspect of a being that itself may be observed in different respects at a given time and that may appear in a different respect at the same time or at a different time, but only as a dependent aspect of the thing whose property it is.[16] "Time" involves being simultaneous with, prior to, and

15. Aristotle, *Metaphysics* IV.1005b 19–20. Though there are ways of expressing the principle in symbolic logic, the Aristotelian formulation has the advantage of being immediately linked to real beings.

16. I am aware of and accept Heidegger's critique of Aristotelian "ousiology" as not properly applicable to *Dasein* in relation to other things and to things as available in the lifeworld. As I understand it, *ousia* has to be abstracted from the network of relations tied to the care-structure of the human being and so is not "primordial." Nonetheless, the "de-worlded" thing still presents itself in a "correct" manner, though it has "lost its being." This is linked to Heidegger's statement that the overcoming of metaphysics is not its elimination but the deeper situating of it in its ground in *Dasein's* life-world. Whatever statements are made about this level are such that they have to cohere with other statements at that level *and* with what is correct at the level of the *Vorhanden*. Heidegger himself says, "For metaphysics overcome in this way does not disappear. It returns transformed and remains in dominance as the continuing difference of Being and beings"; Heidegger, *The End of Philosophy*, trans. Joan Stambaugh (New York: Harper and Row, 1973), 85.

after. "And" as well as "both" involve joining several items, the latter referring to "two." And each of the terms involves "sameness" in the empirical differences from which they abstract and a "difference" from both the empirical instances and other terms compatible with a generic sameness. *Being, becoming and fixity, sameness and difference,*[17] *unity and multiplicity, possibility and impossibility, negation and affirmation, thing and property, joining and separating, past-present-future, appearance and reality* (*part and whole*), and *awareness* in all the aspects operative within it: these various concepts are involved in recognizing and being able to formulate the principle of noncontradiction.

Not only the principle but also the recognitions involved in formulating it were first rendered explicit in Aristotle. They remain the basis for all rational thought and the languages used to express it. Even Kant, with his restriction of knowing to the phenomenal order, nonetheless claimed at least the *thinkability* of the noumenal, with the understanding that whatever is thought about has to be noncontradictory in itself and with everything knowable in the phenomenal order.[18]

GRANTED THAT our mind originally is emptily oriented toward the Whole of what is via the notion of Being, the question is how that emptiness is to be filled in. There is a dialectic of positions in the history of philosophy that turns on the question: is Being a genus? Thinking of it as the highest genus, one would abstract from the Aristotelian categories of entity (*ousia*) and properties (*symbebekota*) as being-in-itself and being-in-virtue-of-another respectively.[19] Empirically, color does not exist in itself but only

17. In the *Sophist* (254b), Plato identifies these as the five great kinds overarching all forms and governing their weaving together.

18. Kant, *Critique of Pure Reason*, A148/B188 to A153/B193, 188–91.

19. For whatever reason, the translation of this as "accidents" is most mis-

as a feature of a colored thing. The concept of Being as abstracted from those two modes—being-in-itself and being-in-virtue-of-another—would be the emptiest of all concepts, as Nietzsche noted. Duns Scotus claimed it had one characteristic that covered univocally all beings, from the lowest of creatures to God: being-outside-of-nothing.[20] Hegel reserved that for the concept of Determinate Being (*Dasein*) or the plurality of beings that was preceded by Becoming as the first concrete concept, a synthesis of Being and Non-Being, both equally empty.[21] Such a synthesis occurred for Plato in *Genesis* or Becoming as a mixture of being and non-being.[22]

Parmenides was the first to isolate the notion of Being and proclaim its identity with thought. In so doing, he implicitly enunciated for the first time the principle of noncontradiction involved. He developed his understanding of Being from the seemingly self-evident declaration given him by a goddess: "It-is and It-is-not is not" or "There is being but no non-being."[23] He goes on to ask the readers to "test by *logos* the truth of what I say." *Logos* typically meant "word," but here it is used to designate for the first time "logic" in the sense of deduction governed by non-contradiction. To claim that non-being *is* is to posit a contradiction. So understood, the principle excludes the two overarching features of everything in our usual experience—

leading. It is no accident that I have brain or a heart, even that I have weight, though the amount may vary over the years. "Properties" works better, since it includes what is essential and what is accidental. Lonergan helpfully translates *symbebekota* as "conjugates," based on an exact Latin parallel to the Greek, "what is co-yoked" with the thing; Lonergan, *Insight*, 437.

20. John Duns Scotus, *Philosophical Writings* (New York: Bobbs-Merrill, 1962), 3ff.

21. Hegel, *Science of Logic*, 105ff.

22. Plato, *Republic* V.478e.

23. G. S. Kirk and J. E. Raven, eds., *Pre-Socratic Philosophers* (Cambridge: Cambridge University Press, 1966), frag. 2 (344), 269.

namely, plurality and change. Regarding *plurality*, as Spinoza later observed, "All determination is negation"; each determinate thing is *not* every other.[24] And, since every extended thing has a plurality of aspects, for Parmenides Being itself must be unextended, neither a one among many nor a one composed of many aspects.[25] It could not therefore be spatial, since space involves a plurality of possible positions. Second, *change* involves things not being what they were and not being what they are going to be—that is, change involves the non-being of past and future and the nontemporally extended, ever-moving Now. So Being excludes both space and time. And, as Parmenides's follower Melissus observed, Being has to be unlimited because there is nothing outside to limit it.[26] So Being considered apart from all non-being or Being as a nature beyond all beings is *un*divided, *un*changing, *beyond* space and time, and *in*finite. This may be seen as the first instance of a negative ontotheology (to employ Heidegger's expression), a determination of the nature of the divine through developing the logic of the notion of Being.[27] Its central problem lay rather obviously in the unintelligibility, because of the contradictory character, of anything outside pure Being. This observation will be taken up by Hegel.

Plato, in the *Sophist*, pointed out that it is through the notion of Being that, as the *Theaetetus* has it, the philosopher has his eyes fixed on the character of the Whole.[28] He announced that he had "to lay patricidal hands on father Parmenides" by distinguishing between absolute non-being and the relative non-being

24. Baruch Spinoza, Letter to Jarig Jelles, June 2, 1674, in Spinoza, *Correspondence*, trans. A. Wolf (Whitefish, Mont.: Kessinger, 2005), 369–70.

25. Melissus drew this conclusion; see "Melissus of Samos," in Kirk and Raven, *Pre-Socratic Philosophers*, frag. 391, 302.

26. Ibid., frag. 3–6 (382–85), 299.

27. Heidegger, *Identity and Difference*, trans. Joan Stambaugh (New York: Harper and Row, 1967), 54ff.

28. Plato, *Sophist* 254A; Plato, *Theaetetus*, 174A and 175a.

of *difference* (*thateron*).[29] Difference would apply both to the region of spatiotemporal things, according to Parmenides's analysis mixtures of being and non-being, and to the realm of Forms that stood beyond the negativity involved in spatiality and temporality, but that were constituted, through the notion of the relative non-being of difference, as (to employ de Saussure's expression), "a system of differences," but also as a system of identities in their differences, specifically and generically.[30]

Plato went on to note that both Becoming (*kinesis*) and the fixity of Forms (*stasis*) *are*, so being cannot be identical with either under penalty of the other not being. However, though not identical with, they both participate in, being. They are both the same in being and different in mode of being. Plato goes further to observe that Difference is both different from Sameness and participates in it by being the same as itself, so Sameness is different from Difference and is itself the same as itself: Sameness and Difference are coimplicated.[31] But the separation of the Forms from their individual instances, the so-called heavenly character of the Forms, is brought down to earth in the *Parmenides* when Plato has Parmenides say, no Forms, no language.[32] Forms are the intelligible ingredients in language embedded in the flow that comes from our mouths in speech.[33]

In the *Sophist* Plato also has the Eleatic Stranger claim that "the definition of being is the capacity to act and to be acted upon"[34]—a protean definition that sees the individual being not as an isolated thing, but as systematically related through its powers to all that it can act upon and that can act upon it. To be an

29. Plato, *Sophist* 257A–259D.

30. Ferdinand de Saussure, *Course in General Linguistics*, trans. Roy Harris (Chicago: Open Court, 1972).

31. Plato, *Sophist* 259A.

32. Plato, *Parmenides* 135C.

33. Plato, *Theaetetus* 208C.

34. Plato, *Sophist* 247E.

individual is to be a cluster of powers as *universally* oriented toward the *kinds* of things and aspects correlative to the powers. As Leibniz saw, each thing has to have within it its compossibility with everything else.[35] The task of experimental science would be to submit each individual of a certain type to tests that show what as a matter of fact its typical cluster of powers is. This allows the scientist to know, in an increasingly refined manner of knowing, the kinds of things and powers in their factual interrelations.

Passive capacities allow for an indeterminate but strictly circumcised way of being acted upon. That matter is extended and resistant allows for manipulation. So a tree is potentially a chair that is intelligible in terms of its manipulability in relation to the form that is given to it from without.

In the history of Platonism, Plato's declaration in the *Republic* that the Good is *epekeina tes ousias* has been regularly but incorrectly translated as "beyond being" and has generated a tradition of nonsense in claiming that God *is* beyond being.[36] I say "nonsense" because to say something *is* beyond being is to position it *within* being understood as *what is the case*. I say "incorrectly translated" because Plato also refers to the Good in the same dialogue as *phanotaton tou ontos*, as the most manifest (aspect) *of being*,[37] and he identifies *ousia* as the realm of Forms correlate to *nous*.[38] *To on* covers both *ousia* and *genesis* or becoming and the Good as well. I suggest that the correlate that is beyond *nous* is eros as a relation of the whole person to the absolute End and Source.

35. Gottfried Wilhelm Leibniz, *Monadology*, in *Leibniz: The Monadology and Other Philosophical Writings*, ed. Robert Latta (London: Oxford University Press, 1951), §§56 and 57, 248.

36. Plato, *Republic* VI.509B.

37. Ibid., V.478E.

38. Ibid., VI.509B.

For *Aristotle*, the question of questions is, What is Being?[39] In his view, being cannot be a generic concept, since genera are abstracted from the conceptual content that falls under their domain as the species are abstracted from their instances: the higher and more extensive the genus, the greater the abstractness. For Aristotle, being is not a genus because it includes all that from which one is abstracted to get the hierarchy of species and genera of progressively wider extension.[40] Being is the concrete totality of things and forms. However, the nature of Being is most fully realized in the highest Form, *Noesis Noeseos*, Knowing of Knowing or Self-Thinking Thought.[41] It is inferred and projected to the top of the hierarchy of the ways in which form rises above the matter it articulates. Form does so in living things where it dominates by organizing matter. In cognitive agents, awareness, rising above matter, involves form-possession without the matter of its object. But in animal awareness, cognition involves possessing form with the *conditions of matter* (individuality, change, spatiotemporal location, self-opaqueness), while in intellectual cognition, the form of intellect (as "in a way, all things") transcends the here and now and is thus able to attend to the form of the known as universal, apart from the conditions of matter.[42] For Aristotle, it is in the human intellect, in imitation of Self-Thinking Thought, that Plato's Forms have their locus.[43]

Aquinas assimilated Aristotle but went considerably beyond him. For Aquinas, Being as *ens* was *esse habens*, that which has existence.[44] This describes, first of all, *ens commune* as object

39. Aristotle, *Metaphysics* VII.1028b.5.

40. Ibid., III.998b.22.

41. Ibid., XII.1072b.20.

42. Aristotle, *On the Soul* III.424a.17–430a.7.

43. Ibid., III.429a.27.

44. Aquinas, *Truth* I.1, in *On Truth*, trans. Robert Mulligan, vol. 1 (Indianapolis: Hackett, 1994).

of human knowing. Aquinas distinguishes *esse* or "to be" from what possesses it—namely, what something is or its *essence*. Essence limits the degree that things are. In a Scotist understanding, that makes no sense: to be is like pregnancy; it either is or is not. In Aquinas's understanding, following Aristotle, being is found in the whole collectively and in each thing taken distributively. There is, consequently, a hierarchy of modes of participating in being, depending upon the degree to which form rises above matter and the degree in which potency comes into actuality—that is, the way in which a thing becomes what it was meant to be.

The notion of Being is without limit, since we can question beyond any putative limit. This poses a new question: how can finite being be at all? It is precisely the *excessus* of the notion of Being beyond what is common to all finite beings that leads to the question of the possibility of absolutely Infinite Being. The limitation of *esse* by essence points to a Ground in *Ipsum Esse Subsistens*, subsistent Existence itself *or Esse* as a nature.[45] By reason of the functioning of the notion of Being, every intellectual being in knowing implicitly knows God, but not as God.[46] That implicit knowledge is the underlying orientation toward the Whole of Being.

In this analysis, there is a distinction between the *concept* of Being and the *judgment* of being that concerns *esse* or existence that is always individual.[47] The analysis of concepts does not yet touch the existence of individuals that fall under the concept. For Aquinas the space of divine omniscience is the realm of pure *Esse* participated in by beings (*entia*) that have *esse* in a finite mode. The *ens* that first occurs within intellectual awareness as its proper object still points beyond itself to *esse* as a na-

45. *ST* I, q. 3, a. 4.
46. *ST* I, q. 2, a. 1, ad 1.
47. *ST* I, q. 85, a. 1 and 2.

ture. This escapes Heidegger's critique of ontotheology, or the attempt to understand the divine through an analysis of the logic of the notion of Being. Aquinas escapes the critique because *esse* is transconceptual and always individual, reached not by means of the universal concept but by means of existential judgment.

For *Spinoza*, philosophers so far have begun with abstractions, whether sensory or intellectual. He proposes to begin, in the Aristotelian line, with Being as the most concrete concept, including as it does all that is.[48] But everything that we would call a being does not truly exist "in itself" (the definition of thing or substance), since, in the line of Aquinas, it is unintelligible without its ground in the encompassing single Substance, *Deus sive Natura*, which does not exist in virtue of anything more encompassing.[49] Things are modes or aspects of that single Substance. Spinoza understands the system of things as *Natura naturata*, nature being "natured" or continually produced by *Natura naturans* or nature as an intelligible system.[50]

Hegel claimed that to think properly is to think like Spinoza—that is, to see everything rooted in the overarching reality of the Divine. But, by reason of human freedom rooted in being projected toward being as a Whole via the notion of Being, human existence shows that rationally free subjects are not simply distinct from, but stand over against the ground within which they are rooted. So Substance has to become Subject. Free subjectivity has to be included in the overarching Substance.[51]

48. Spinoza, *On the Improvement of the Understanding*, trans. R. H. M. Elwes (New York: Dover, 1955), 15 and 29.

49. Spinoza, *Ethics*, trans. R. H. M. Elwes (New York: Dover, 1955), I, Prop. XV–XVI, 55–9.

50. Ibid., I.XXIX, note.

51. Hegel, *Phenomenology of Spirit*, trans. A. V. Miller (London: Oxford University Press, 1977), §17.

Given our ability to question beyond any putative limit, the question arises of how being can be limited to any finite mode. For Hegel, the limitation involved is one way of seeing that "non-being is," for limited being is a negation of the anticipated infinitude of being and is thus a "contradiction" that must be resolved.[52] The notion of contradiction here follows Parmenides: any way in which even relative non-being is contradicts pure Being. Thus appetite as a lack is a form of non-being that is, which is a contradiction calling for resolution in eating.[53] For Hegel, only an absolutely Infinite Being giving being to finite being resolves the contradiction of finitude itself. The climax of Hegel's own System lies in Aristotle's *Noesis Noeseos*, the highest instance of Being, now, as in Aquinas, rendered infinite and as Ground of all that is.[54]

Now, Hegel takes the concept of Being as a genus, abstracted from the Aristotelian categories of substance or thing as being-in-itself and accidents or properties as being-in-virtue-of-another. The Parmenidean set of negations (unchanging, undivided, infinite, nonspatial, and nontemporal) gives us an empty rather than a full Being.[55] But it is just that emptiness that impels thought beyond the concept of Being, initially as apart from all filling, toward Absolute Knowing as complete systematic conceptual filling. For Hegel, the empty notion of Being develops into the notion of Becoming, presented initially by Plato

52. Hegel, *Hegel's Science of Logic*, trans. A.V. Miller (Oxford: Oxford University Press, 1976), 136ff.

53. Hegel, *Hegel's Philosophy of Mind*, trans. Michael Inwood (Oxford: Oxford University Press, 2007), §426.

54. The last line of Hegel's treatment of Spirit as the end of the System is a quotation in Greek of Aristotle's description of Self-Thinking Thought; *Hegel's Philosophy of Mind*, §577.

55. Hegel, *Science of Logic*, 94. See also his treatment of Parmenides in Hegel, *Lectures on the History of Philosophy*, trans. E. S. Haldane (Lincoln: University of Nebraska Press, 1995), 1:249–60.

as a synthesis of being and non-being.[56] There follows a dialectical unfolding that progresses until the essential hidden potency of the notion of Being is shown by the generation of the interlocking set of concepts that displays the form of the cosmos that makes rational existence and flourishing possible.

Though claiming through this System to "think the thoughts of God before creation,"[57] Hegel's Absolute Knowing is not omniscience. It is much more "modest": it displays the interlocking set of ontological, cosmic, anthropological, personal, historical, and institutional conditions for the possibility of rational existence and flourishing. It is a framework analysis that situates rather than dictates the actual choices made by existing subjects. Hegel retrieves the ontological argument for God's existence by pointing out that the human being, as oriented toward the fullness of Being, is *itself* the ontological argument, the transition between concept and existence.[58]

However, in addition to Absolute Knowing, God's omniscience appears in Hegel's philosophy of religion, but is not further developed.[59] It would be an *existential* transcendence, *beyond* the Absolute Knowing of the conceptual system—a transcendence *beyond*, as correlate to actual, individual, finite existence *within*, the framework of the System. The conceptual analyses in Parmenides, Spinoza, and Hegel would then require supplementation through the judgment of existence. This is where Hegel could join with Aquinas.

Heidegger claimed that with Hegel metaphysics reached a kind of culmination. Heidegger's criticism of ontotheology is directed against Hegel's conception and against the way of tak-

56. Hegel, *Science of Logic*, 105–6.

57. Ibid., 50.

58. Hegel, *Lectures on the Philosophy of Religion*, ed. and co-trans. Peter Hodgson (Berkeley: University of California Press, 1988), 167.

59. Ibid., 419, 474.

ing Parmenides's conception we have sketched out. The question of Being is central to human being as *Dasein*, as the locus of the showing of Being among beings.[60] Apart from the limited range of the other sciences, metaphysics aimed at investigating being as being, understood as things and their most universal principles. Heidegger claims that, in this focus, metaphysics has issued in "the forgottenness of Being."[61] Using Descartes's image of the Tree of Knowledge, where metaphysics functioned as the roots, Heidegger asked, What is the soil in which the roots are planted and from which the tree derives its nourishment? What is the ground of metaphysics?[62] Basically, it is the life-world of a people articulated by the poets.

Heidegger's orienting experience was of "startled dismay" (*Erschrecken*) at the forgottenness of Being in and through this tradition.[63] Things, he said, "have lost their being" and have become data or "standing reserve" for our projects.[64] Here he draws upon the proclamation of Stephan George: "Where word breaks off no thing can be."[65] Gabriel Marcel had the same type of experience: he said that things have lost their "ontological weight."[66] The possibility of nearness, proclaimed in poetry, is lost in placing things at a distance for intellectual and practical

60. Heidegger, *Being and Time*, trans. Joan Stambaugh (Albany: SUNY Press, 1996), 11.

61. Heidegger, "The Way Back into the Ground of Metaphysics," in *Pathmarks*, ed. William McNeill (Cambridge: Cambridge University Press, 1998), 277–90.

62. Ibid., 277.

63. Heidegger, *Contributions to Philosophy: On Enowning*, trans. Parvis Emad and Kenneth Maly (Bloomington: Indiana University Press, 2000), 11.

64. Heidegger, *The Question Concerning Technology and Other Essays*, trans. William Lovitt (New York: Harper and Row, 1977), 17–20.

65. Heidegger, "Words," in *On the Way to Language*, trans. Peter Hertz (New York: Harper and Row, 1971), 139–56.

66. Gabriel Marcel, *Existential Background of Human Dignity* (Cambridge, Mass.: Harvard University Press, 1963), 74–75.

mastery. As the element of fish is water and of birds, the air, the element of humanness is in the nearness of things and the Mystery in which they are rooted.[67]

What has been forgotten in the metaphysical pursuit is truth as *a-letheia* that opens the horizon for truth as correspondence. *A-letheia* points to the unconcealed that rises up from the *Lethe*, the essential concealment out of which things are manifest, not only in their individual fullness but in their essential interrelatedness within the Whole. The *Lethe* is the Mystery of Being. For Heidegger, conceptual being is not Being itself; its showing, as any showing, always involves a concealing. Heidegger attempts to get back to the ground of metaphysics, not in the concept, but in the *Sinn des Seins*, the sense of Being.[68]

Pondering "the ground of metaphysics" is his own way of taking the notion of Being. It is correlate to a transintellectual relation of dwelling, of the drawing near of beings. In dwelling we find a correlate to the notion of the human heart in *Stimmung*, the attunement of the human being. Dwelling with one's whole being is represented traditionally, but not exploited philosophically, by the notion of the heart.[69] Heidegger goes back to reread Parmenides. *Noein* as thinking, he says, "is the apprehension that has man" being appropriated by the Mystery of Being.[70] I am surprised that he did not make use of the opening

67. Heidegger, "Letter on Humanism," trans. Frank Capuzzi, J. Glenn Gray, in *Martin Heidegger: Basic Writings*, ed. David F. Krell (New York: Harper and Row, 1977), 195.

68. Heidegger, "Way Back into the Ground of Metaphysics," *Pathmarks*, 217.

69. See Wood, introduction to *Phenomenology of Feeling: An Essay on the Phenomena of the Heart*, by Stephan Strasser, trans. Robert E. Wood, preface to the English translation by Paul Ricoeur (New York: Humanities Press, 1977); see Wood, "The Heart in Heidegger's Thought," *Continental Philosophy Review* (2015): 445–62.

70. Heidegger, *Introduction to Metaphysics*, trans. Gregory Fried and Richard Polt (New Haven: Yale University Press, 2000), 150.

line of Parmenides's work: "The steeds that bore me took me as far as my *heart* (*thumos*) would desire." In his exploration of the poets, Heidegger concentrates, among other things, upon Rilke's notion of the heart as correlate to "the globe of being."[71]

I would contend that the heart lies at the heart of Heidegger's thought.[72] Heart is that center of being attracted, repelled, or rendered neutral by what one confronts. It is the default mode that guides our operations. It is that by which we are mastered, taken outside ourselves, touched, and possibly transformed. It is through the heart that we inhabit our life-world where things are set at a distance or are able to draw near. Thinking that is focused on recovering the forgotten Mystery of Being is appreciative thinking, thinking as thanking, meditative thinking that evokes presences.[73]

Heidegger further claims—without to my knowledge having pursued it further—that the thought of Being he is exploring may contribute to a further understanding of what Aquinas meant by *esse*.[74] Along these lines, what Josef Pieper called "the silence of St. Thomas" is the mystical rootage of his own heart in the presence of the divine *Esse*.[75] For Aquinas, prayer is the unfolding of the mind before such Presence that may appear more or less intensely as an arresting Presence or recede, as *its* default mode, into silence.

Derrida, coming out of Heidegger, argued against Benve-

71. Heidegger, "What Are Poets For?," in *Poetry, Language, and Thought*, trans. Albert Hofstadter (New York: Harper and Row, 1971), 124.

72. See Wood, "The Heart in Heidegger's Thought," *Continental Philosophy Review* (2015): 445–62.

73. Heidegger, *What Is Called Thinking?*, trans. Fred Weick and J. Glenn Gray (New York: Harper, 1968), Part II, Lecture III, 138ff.

74. Heidegger, "Way Back into the Ground of Metaphysics," in *Pathmarks*, 210.

75. Josef Pieper, *The Silence of St. Thomas*, trans. Daniel O'Connor (New York: Pantheon, 1957).

niste's attempt to reduce Greek ontology to Greek grammar and defended the notion of Being as transcultural ground of the categories.[76] But further, for Derrida, the ontological difference between Being and beings is grounded in what he calls *différance*, the system of differences that is the network of language and the deferral of full presence in our relation to things.[77] Carrying out his deconstructive project, he aimed "to send metaphysics packing,"[78] which he falsely claimed is Heidegger's aim. That is something Heidegger explicitly denies: what is required is a *Zuspiel*, a counterpoint to the tradition of metaphysics in relation to a second beginning that thinks the ground of metaphysics. Even for Heidegger, metaphysics is not a mistake but an essential call of human being.[79]

FINALLY, LET US circle back to our initial three propositions. The notion of Being makes possible both apprehension of the universal and freedom of choice, according to Aristotle the two distinctive characteristics of the human being whose mind "is, in a way, all things."[80] For Aristotle, such a being is *zoion politikon*, necessary inhabitant of a tradition reciprocally tied to being *zoion logon echon*, having the capacity for rationality actualizable only through linguistically mediated tradition.[81] Boethius's translation of the latter into Latin as *animal rationale* flattens out the Greek meaning, for *logos* has, in the case in

76. Jacques Derrida, "The Supplement of the Copula," in *Margins of Philosophy*, trans. Alan Bass (Chicago: University of Chicago Press, 1986), 182ff.

77. Derrida, "Différance," in *Margins of Philosophy*, 7–8.

78. Derrida, *On Spirit*, trans. Geoffrey Bennington and Rachel Bowlby (Chicago: University of Chicago Press, 1989), 75.

79. Heidegger, "Way Back into the Ground of Metaphysics," in *Pathmarks*, 209.

80. Aristotle, *On the Soul* III.8.

81. Aristotle, *Politics* I.1.1253a.9.

question, two different but related meanings: it means *discourse* as well as the *rational capacity* for discourse. That would make the two Greek definitions closely related, for the *zoion politikon* is so by reason of the *logos* of language that mediates the tradition. As in Hegel, the universal structure of being a human subject is only actualizable through the objectifications of other subjects that endure, beyond their inevitable demise, in the form of institutions, beginning with the institution of language whose sensory immediacy we have focused upon in the beginning.[82]

So the notion of Being that first arises in the intellect grounds our ability to understand. Its establishing primordial distance from all finitude also grounds our ability to choose to pursue whatever we chose consonant with the available possibilities. Here we have chosen to examine the proposition that began our inquiry: "First to occur in intellectual awareness is the notion of Being." In the process, we have also illustrated our third proposition: that reflective knowing involves "the complete return of the intellectual subject within itself" or, in Aquinas's Latin, the *reditio completa subiecti in seipsum*, making manifest the essential parameters involved in being such a subject: sensory encounter, reference to the Whole of what is via the notion of Being, reference to space and time as encompassing wholes, linguistic mediation, transcendence of ethnic peculiarities involved in language and practice, responsible choice within and beyond those peculiarities, and the personal centrality of the heart. The return is "within" the subjectivity of awareness insofar as all its operations, even those directed to what lies without, are simultaneously within the field of awareness on the subject side. The return makes us aware reflectively of the parameters involved in the total field of our awareness, aware of ourselves

82. Hegel, *Phenomenology of Spirit*, §171.

as the subject for which everything else makes its direct or indirect appearance.

The claim that *Primum quod cadit in intellectu est ens*, "that which first arises within intellectual awareness is the notion of Being," identifies that which makes us the *animal rationale* who develops through linguistic communication and who realizes the full parameters of its own being in the *reditio completa subjecti in seipsum*, aware of the essential parameters of its own functioning such as we have laid out in this chapter. It is, as well, free to determine itself. The three initial propositions are not simply claims made by Thomas Aquinas, but deep truths involved in our being human.

3 The Undeconstructible Foundations of Human Existence

On the Magnetic Bipolarity of Human Awareness

DECONSTRUCTION and anti-foundationalism play a large role in contemporary philosophy.[1] One might deconstruct various aspects of our life-world and the texts we employ in interpreting that world; and one might reject the way foundations were intended to work in the *more geometrico* procedures of Descartes and Spinoza;[2] but the practice of deconstruction and the ar-

1. Jacques Derrida is the father of deconstruction and a prominent anti-foundationalist; see, for example, *Of Grammatology*, trans. Gayatri Chakravorti Spivak (Baltimore: Johns Hopkins University Press, 1976). Michael Foucault historicized all comprehensive modes of thought; see, for example, *The Archaeology of Knowledge*, trans. A. M. Sheridan Smith (New York: Pantheon, 1972).

2. René Descartes is the father of modern foundationalism, finding the basis for thought in the self-presence of the *cogito*; see Descartes, *Meditations, Objections, and Replies*, trans. Roger Ariew and Donald Cress (Indianapolis: Hackett, 2006). Benedict Spinoza aimed at a system of the Whole proceeding *more*

guments launched against foundationalism presuppose a foundation, itself undeconstructible, in the structure of the field of human experience. That is what I intend to explore in this chapter. I will work with what is most exterior to what is most interior to our existence. And what I claim is that the kinds of evidences I will examine are presupposed as foundation in all our wakeful activity.

While animal awareness is monopolar, human awareness is, like a magnet, bipolar. As Dewey noted, guided primarily by seeing, an animal lives outside its own skin in relation to what appears sensorily in its environment and in function of biological needs.[3] As we noted in chapter 1, since people began thinking about seeing on the basis of neural research, the natural tendency is to think of seeing as occurring inside the brain at the end of a causal chain that begins with radiating light sources that are partially absorbed, partially reflected by bodies. The reflections pass through the pupils and are inverted by the ocular lenses to affect the rods and cones of the retina, setting off electrical impulses that pass, via a sequence of electrochemical switches along the optic nerve, to the visual cortex in the back of the brain where color is produced. The claim regularly made is that what the seer really sees is a peculiar colored object inside his or her brain. But this contradicts the evidence that is supposed to support this claim, for the whole sequence is viewed *outside* the neurophysiologist's brain. He, like other animal beings, lives outside his or her skin in the environment.[4]

This living outside one's skin is one of the several eidetic

---

geometrico, beginning with fundamental principles and deducing every fundamental type from that; Spinoza, *Ethica*, trans. R. H. M. Elwes (New York: Dover, 1955).

3. John Dewey, *Art as Experience* (New York: Putnam, 1958), 58; Dewey, *Experience and Nature* (New York: Dover, 1958), 282.

4. See chapter 1, on seeing.

features that constitute the enduring framework of seeing, features verified every time we open our eyes. Among other things, what seeing involves is the illusion of emptiness in phenomenal space. Along with air molecules and macrolevel sounds, actual space is full of all the irradiations across the electromagnetic spectrum. Most of that is filtered off by the structure of our optical system to allow an animal to see prey or predator approaching at a distance. The senses as such present us with modes of appearance determined by the interplay of real things and real perceivers to satisfy real needs.

Things appear in the visual field within the horizon as the limit to that field. Orientation toward the horizon involves perspectival distortion, since things phenomenally shrink as they recede from our bodily point of view. Horizon, perspective, and phenomenally empty space are neither wholly subjective nor wholly objective: they are a result of a meeting of environmentally given things and a bodily situated, appetitively driven awareness. As Aristotle has it, "The sensible in act is the sense power in act."[5] There is a cognitive identity between awareness and its object, what Hegel will call "identity-in-difference."[6] This type of relation holds also for color itself, inhering necessarily in extension, as the aspect revealed by seeing. Color is not *in* us or *on* things outside us: as Plato noted, it emerges in the between, in the meeting between seeing and the seen.[7]

Perspective wasn't made focal until Brunelleschi and Alberti discovered it during the Renaissance.[8] The reason it had not

5. Aristotle, *On the Soul* II.425b.27, trans. J. A. Smith, in *Complete Works*, ed. Jonathan Barnes, vol. 1 (Princeton: Bollinger, 1984).

6. G. W. F. Hegel, *Hegel's Philosophy of Mind*, trans. Michael Inwood (Oxford: Oxford University Press, 2007), §566–70, 299–301.

7. Plato, *Theaetetus* 153D.

8. See Hubert Damisch, *The Origin of Perspective* (Cambridge, Mass.: MIT Press, 1994).

been noticed before was that our neuropsychological sorting system discounts the distortions because we have already seen things from various angles. Our sensory system also integrates the seen with the heard, smelled, touched, and tasted to respond to the types of things that reveal and conceal themselves through our senses.[9] Our system develops its learned universal orientations on the basis of its native powers. (We will return to this point later.) An animal's neurosystem learns how to respond typically to typical instances of its prey in the typical evasive moves it makes.

What is crucial is that the awareness involved must be a kind of self-presence, because what is manifest is manifest as *other than* that self-awareness: manifest otherness entails manifest identity other than that in which the other is present. This again is a most important insight. A photoelectric cell hooked up with a computer cannot see because it has no self-presence, and that because it has no appetites and thus no feeling. When the "consciousness switch" in the reticular activating system located in the stem of the brain is "on," feeling suffuses the organic system. Feeling is the nonreflective self-presence of an animal being. The self-awareness, governed by felt appetites, makes possible the manifestation to it of what is other. The integrating work of sensibility is for the sake of animal flourishing through serving the appetites. That is indicated by the feeling of pleasure when the appetites are fulfilled, pain when they are not, or when one is severely impacted in a negative way by the environment.

What appears in such a world are always individual actualities, never universals or potentialities, even though the latter, as native and learned powers, are the ground of what an animal is

9. See Maurice Merleau-Ponty, *Phenomenology of Perception*, trans. Donald Landes (London: Routledge, 2012), 100–106.

able to do. Each native potentiality is a universal orientation in an individual entity toward *all* the instances of its formal object, the aspect of things that the power reveals, as color is revealed by seeing. Natural and acquired potentialities existentially undercut the distinction between the individuality presented sensorily and the power's own concrete universality, as well as the universality of the type of object correlative to each of the powers.

I take this to be a most important insight. Ordinarily we think, if we think about it at all, that the universal terms we use to characterize things and their properties belong to us, while individuality characterizes the things about which we speak in universal terms. The problem is how to bridge the gap. As a matter of fact, things themselves already bridge the gap because their powers, both native and acquired, are universal orientations toward *all* the individuals correlative to the power that are, in turn, instances of the *type* correlative to the power. Thus seeing as a native capacity is oriented toward all the seeable but is only actualized by individual things seen.

In terms of learned capacities, an adult cheetah has learned how to respond to the differing kinds of moves his kind of prey might make. Carpentry, as a distinctively human learned capacity, is oriented toward all the *kinds* of objects, correlative to the skill level of the carpenter, upon which it can be employed. Capacities, native and learned, require natural and manufactured *kinds* of individuals both on the part of the possessor of the capacities and on the part of the aspects of things correlative to those capacities.

Plato generalized that notion in his *Sophist*, proposing that a thing is a cluster of powers of acting and being acted upon.[10] As such, it cannot adequately be considered by itself, for it requires an environment of kinds of things and their specif-

10. Plato, *Sophist* 247E.

ic kinds of aspects corresponding to its powers and those of the gamut of environing things. Science puts things to the test to find out what their kinds of powers actually are.

Thus far the organically based sensory pole. The undeconstructible claim I am making involves the factors we have examined as always operative in normal wakeful life.

ANIMAL AWARENESS, we said, is monopolar, its awareness governed by organic need that requires individuals of specific kinds here and now manifest in the environment. Now what allows us to recognize the universal kinds and the universally oriented powers as such is the *bipolarity* of the field of human experience. Distinctively human awareness is oriented beyond the here and now by reason of being oriented toward *all* that is, individual or universal, finite or infinite, temporal or eternal. It is so oriented by reason of the notion of Being that first arises within the mind to make it a mind. As Aristotle noted, the mind is, "in a way, all things."[11] It is by way of being oriented via the notion of Being toward the totality of what is.

We said that orientation via the notion of Being in an initially empty way toward the absolute totality makes the mind be a mind. In our case, this involves orientation toward the Whole that is our native abode, the Whole accessible by way of our sensorily given starting point: the totality of the spatiotemporal universe. As Kant noted, this involves the forms of space and time as all-encompassing a priori, forms of sensibility given by immediate intuition, to be concretely filled by experience and its inferential extensions.[12] It is this a priori orientation that

11. Aristotle, *On the Soul* III.425b27.

12. Immanuel Kant, *Critique of Pure Reason*, trans. Norman Kemp Smith (New York: St. Martin's, 1969), A 22/B 37, 67ff.

grounds our ability to recognize kinds as kinds, to abstract the kind from the individual instance, and to see that the kind applies *anytime* and *anyplace* instances of the kind are met. *Anytime* and *anyplace* have therefore to be emptily pregiven to the particular times and places we come to know.

There is another aspect to this. We are able so to operate, being able to abstract, because our rationality has always already taken over our sensibility through the creation of language, the sensory and communal vehicle of thought. As Hegel noted, reason can only develop by laying down its tracks in the sensory realm through the creation of language.[13] Human rationality, which operates at the level of the universal, involves, by its own nature, incarnating itself in the sensory through language. Outside of proper names, the words we use are stand-ins for our apprehension of universals. We do not typically notice this because both words and the concepts for which they function as vehicle are like glasses: we attend to individual things *through* them without noticing them—unless, as we are now doing, we pay explicit reflective attention to them.

There is a further aspect: we do not give language to ourselves; we were inducted into it by our primary caregivers, and it is further developed as we communicate with each other. Hence, not only do our senses reveal to us features of what is "out there," beyond our private sensations, the language we use to describe them places us in a space of common meaning that allows us to speak to others and to ourselves, even about the features of our privacy. A psychologist once said that we are all trapped in our subjectivity. I said, I know perfectly well what he means. Privacy is precisely priva tive of the common space of meaning that positions us outside of ourselves in an even more comprehensive way than does our sense experience. We not only live outside

13. Hegel, *Hegel's Philosophy of Mind*, §§456–64.

our skins with what appears in our environment, we live outside our privacy in a communal, linguistically mediated life-world.

So we are spatially situated as organisms, living in the surrounding space as sensorily desirous, attending to the present by reason of past learning and oriented toward the future by appetite. There is a standard development, flourishing, and expiration time frame for each kind of organism. Animals operate out of a lived awareness of temporal and spatial extension: the past of learning enters into the present focused on what appears in function of the anticipation of future satisfaction. Subsuming all that, as human beings we live out of a background awareness of space and time as encompassing wholes, are able to choose our goals, and, by reason of the unlimited orientation of the notion of Being, are able to question even whether there might be something beyond space and time.

With the unrestrictedness of the notion of Being goes the unrestrictedness of the principle that grounds rationality: the principle of noncontradiction. As Aristotle formulated it, "It is impossible for the same attribute at once to belong and not to belong to the same thing and in the same relation."[14] The ability to so formulate presupposes apprehension of a large amount of previous experience that we have explored before, but is worthwhile reviewing.

The formulation of the principle presupposes induction into a linguistic tradition. It presupposes an awareness of distinct *things* that are given as enduring throughout the temporal sensory display of differing *properties* grounded in the thing, playing in tandem with our awareness of ourselves enduring through the variation of experiences with regard to things. This experience involves the awareness of *sameness and difference*: the same thing

14. Aristotle, *Metaphysics* 1005b.19. Though symbolic logic may give a different presentation, Aristotle's formulation arises out of the life-world itself.

enduring through the difference of its properties; the same subject of experience enduring, as it attends to the things around it, through the difference of its own properties. Formulating the principle of noncontradiction presupposes as well our experience of past, present, and future as eidetic features of *time*. Differing aspects of things and our different relations to them furnish the *respects* from which a thing can be considered.

The formulation of the principle of noncontradiction also involves awareness of *possibility* opened up by our awareness of typical modes of display through anticipation and inference. It involves as well awareness of *negation* that we come to know by an awareness of lack, of difference, of hopes and fears and the like. But over and against this, there is a *cannot* involved here, a strong sense of impossibility. In the formulation of the principle of noncontradiction, the notion of *both/and* comes from a comparison of two items. Here it qualifies *possession*, having certain properties. A thing cannot both have and not have the same property at the same time and in the same respect.

Finally, the formulation and the presuppositions of experience it requires involve the identity over time both of the thing involved and of the knower as well as the identity of the meanings involved. Once again, "It is impossible for the same attribute at once to belong and not to belong to the same thing and in the same relation."

So there are several presuppositions, experiential and logical, for the formulation of the principle of noncontradiction. It is that principle that opens the space of common meanings limned by logical structure that is the same for all cultures; it is a structure involving absolute impossibility. Empirical impossibility is a matter of fact, like the impossibility of the dead rising again—which may have been countermanded in some instances. But the impossibility involved in the principle of noncontradiction is not empirical; it is absolute. It sets us outside our privacy in the space of common meaning determined by the absolute necessity of logical structure. That space gets concretized through the development of the culture and of the individual within that logical framework.

There are two important insights connected with the principle of noncontradiction. The first is this: any contingent event—that is, one that need not have happened but did (for example, inserting this paragraph into my chapter here), is such that it can *never* afterward not be true that it occurred. That seems evident, but how can we conceive "never"? Peter Damian claimed that what happens in God's forgiving sin is that, because his power knows no restriction, he erases the past sin, declares that it didn't happen.[15] The nonsense this involves is patent; but the insight that counters this entails the recognition of eternity.

How can we recognize that contingency eventuates in necessity or time in eternity? Whitehead argues for a divine Cosmic Memory that retains the past forever.[16] However, this does not explain *our* recognition of it. It is linked in some way to the unrestricted character of the notion of Being. We always situate whatever we think of within the all-encompassing character of what is the case, including the whole of time emptily intended.

The subsequent question is whether such eternity is that of the temporal process itself. Nietzsche lived out of his love for *tiefe, tiefe Ewigkeit,* which he located in time itself as the Eternal Recurrence of the Same.[17] Aristotle also claimed the eternity of the temporal universe but argued also for eternity beyond time in the

15. Peter Damian, "On Divine Omnipotence," in The Letters of *Peter Damian*, trans. Owen J. Blum, vol. 4 (Washington D.C.: The Catholic University of America Press, 1998).

16. Alfred North Whitehead, *Process and Reality: An Essay in Cosmology* (New York: Harper, 1929), 530.

17. Friedrich Nietzsche, *Thus Spoke Zarathustra: The Portable Nietzsche*, trans. Walter Kaufmann (New York: Viking, 1954), 3:15, 3

case of the Unmoved Mover, the Self-thinking Thought necessary to account for the eternity of a temporal universe, each item of which is able not to be.[18] A key question that follows is whether the eternity of the Unmoved Mover is unchanging endurance through time or transcendence by grounding time itself.

These insights give the basis to the claim made by Aquinas that humans live on the horizon between time and eternity.[19] We cannot collapse eternity into Heideggerian Being-in-Time, even though it is within that temporality that eternity makes its appearance. This is why Scheler claimed, against Husserl, but also, in effect, against Heidegger, that we do not simply live in time, we live *into* time from being *beyond* time.[20]

Returning to our previous considerations, granted the principle of noncontradiction, we have also to grant the insights presupposed in its formulation that remain constant through the multiple ways we experience and think about things. The formulations of deconstructive efforts and of the anti-foundationalism in the current philosophical community presuppose what I have formulated up to this point. Indeed, pointing to the wholeness of each thing within the absolute Totality, the notion of Being that grounds everything distinctively human indicates an excess in things that always transcends whatever hold we have on them. We are constantly called to resituate what our current knowledge displays, but always only on the basis of the structural factors I have ferreted out thus far. As Bernard Lonergan noted, "Everything is subject to revision except the conditions for the possibility of revision."[21]

18. Aristotle, *Metaphysics* 12.1074b.

19. Thomas Aquinas, *Summa contra Gentiles*, vol I, lxl.

20. Max Scheler, *Formalism in Ethics and Non-Material Ethics of Value*, trans. Manfred Frings (Evanston, Ill.: Northwestern University Press, 1973), 385.

21. Bernard Lonergan, *Insight: A Study of Human Understanding* (London: Longmans, Green, 1958), 335–36.

We have previously referred to Derrida's claim, against Benvenista's attempt to reduce ontology to Greek grammar, that the notion of Being is transcultural and thus, in some real sense, foundational.[22] Derrida also argues that our relation to real things is one of *différance*: deferral of full presence within linguistic systems as systems of differences. We are never in the full presence of real things and can never fully carry out their definitions. He in effect follows Husserl's noting of the "internal horizon" of real things.[23] Whatever we know of them we know within our human perspective, which can be ongoingly expanded through scientific research. Aquinas had said, "We do not know the essence of things, even so simple a thing as a fly."[24] Derrida also notes that we define in terms of linguistic systems of differences, so that our relation to real things is one of *differánce*, of deferral of full presence and of location within systems of difference that can vary at differing junctures.[25] Kierkegaard gives a perfect example of multiple readings in his treatment of the Abraham story in *Fear and Trembling*.[26]

Such claims, I think, are true. But they presuppose the analysis of the field of experience we have given, showing the sensory and cultural perspectivity of our access to things. But I would add that the linguistic sounds may be systems of differences, but

22. Jacques Derrida, "The Supplement of the Copula," in *Margins of Philosophy*, trans. Alan Bass (Chicago: University of Chicago Press, 1986), 182ff.

23. Edmund Husserl, *Ideas for a Pure Phenomenology and Phenomenological Philosophy*, trans. Daniel Dahlstrom (Indianapolis: Hackett, 2014), §44, 77.

24. Aquinas, *Commentary on Aristotle's "De Anima,"* trans. Kenelm Foster and Silvester Humphries (South Bend, Ind.: Dumb Ox, 1994), I:1, no. 15; Aquinas, *Disputed Questions on Spiritual Creatures*, trans. Mary Fitzpatrick and John Wellmuth (Milwaukee, Wisc.: Marquette University Press, 1949), 11, ad 3; Aquinas, *On Truth*, trans. Robert Schmitt, 3 vols. (Indianapolis: Hackett, 1994), vol. 4, a 1, ad 8.

25. Derrida, "Différance," In *Margins of Philosophy*, 12–13.

26. Søren Kierkegaard, *Fear and Trembling*, trans. Alasdair Hannay (London: Penguin, 1985).

they are also linked to meanings that can be defined in systems of *samenesses* and differences: species sameness within the differences of individual members of a given species and generic sameness within the differences of species or less comprehensive genera falling under a genus. The generic hierarchy provides an inventory of the kinds of beings that constitute the Cosmos.

WHAT I HAVE EXPLORED thus far is the *cognitive* realm grounded in the operation of the notion of Being. It is that same notion that grounds volitional activity. It is our being directed, beyond the here and now of our dealings with individuals, toward the Totality that gives us primordial distance from any finite determination, as both Aquinas and Hegel—and Sartre in his own way—recognized.[27] Human awareness is an other to every other, turning us over to ourselves to choose what to make out of ourselves. As Hegel put it, the ability to say "I" involves abstracting from all determinate content, providing an ability to relate ourselves freely to any determinate content.[28]

Now "ourselves" in each case does not simply consist of our mental powers; it includes our total range of native powers (metabolic, sentient, and rational) and all of our acquired powers, beginning with language use. Each self consists of an I-Me relation. Each of us as an I has to work with what is given as Me, that is, with everything determinate that I have become at any given moment. This involves three levels, one chosen, two not chosen but given. The first of the latter is one's genetic endowment that sets limits to what one can achieve as an individual. The second nonchosen aspect is the cultural shaping each of

27. Aquinas, *ST* I-II, q. 10, a. 2; G. W. F. Hegel, *Science of Logic*, 583; Jean-Paul Sartre, *Being and Nothingness*, 478.

28. Hegel, *Hegel's Philosophy of Mind*, §381, *Zusatz*, 12.

us has received, beginning with the discipline provided by one's parents and including especially language through which everything else is mediated. In both cases—genes and culture—I experience myself as "thrown," to use Heidegger's expression (though I would suggest rather considering it as a gift that opens up our concrete possibilities).[29] But when the notion of Being first emerges (determination of the time when this occurs in an individual life is an empirical question), I am set at a distance from the genetic and cultural Me so as to enable me to choose for myself: I am oriented toward the future as the realm of the ends I choose to pursue. To use Heidegger's terms again: I am a *thrown project*, oriented primarily to the future, who takes up the present in function of the projection of the future.[30] In a more Hegelian way of putting it, the I is the cutting edge of a multileveled, psychophysical whole: a metabolic, organismic, sentient-appetitive, intellectual-volitional, historically situated, self-formed Me. In this way each of us must be the artist of her or his own life, working with one's uniquely determinate Me. I have to decide at each moment what I am going to do with the currently determinate Me. The I is the cutting edge of a multileveled, psychophysical whole operating within a culture.

To begin with—and all too often to end with—I choose on the basis of my felt proclivities. The history of those choices produces the character I now have that both opens up and limits my real possibilities, carrying me through my daily life. The vision I have of my own possibilities may be quite narrow. One of the functions of liberal education is to make us aware of wider possibilities.

Most important, the mass of this tripartite determinism settles down in the emotive center that a long tradition calls "the

29. Heidegger, *Being and Time*, 164–68.

30. Ibid., 205.

heart." Each of us is a thrown project as a radically individual magnetic field that involves spontaneous attractions and repulsions formed in the dialectic between the individual's heart and the aspects of the environment, natural and cultural, to which one is most basically attracted. The "reasons of the heart" are the default mode for our spontaneous choices. We are each, for the most part, governed by our heart.[31] It is the determinate aspect of the Me that is closest to the I. This is in a way proximate to Hume's contention that reason is the slave of the passions, only our view requires us to judge and potentially change our hearts through our reason: where is my heart? Is it where it ought to be? Where ought it to be?[32]

Think of our basic bipolarity—organic rootage playing in tandem with reference to the Whole—forming a magnet with two poles that create an overall magnetic field with biological attractors and an overall ontological attractor. Cultures establish their own magnetic fields within the overall field of human structure, often with conflicting attractions and repulsions. A human individual comes to exist and operate within these modulated fields, but through choices interplaying with its genetic and cultural formation, each individual comes to establish her or his own field that governs his or her spontaneous, thoughtless reactions. The fundamental issue is how each of us aligns ourselves in relation to the ontological attractors: how we understand and come to dwell in relation to the Whole. The fundamental matter of the heart.

31. Max Scheler, "Ordo Amoris," in *Selected Philosophical Essays*, ed. and trans. David Lachterman (Evanston, Ill.. Northwestern University Press, 1972), 116. See my treatment of this in Robert E. Wood, "Virtues, Values, and the Heart," in *Phenomenology and Virtue Ethics*, ed. Kevin Hemberg and Paul Gyllenhammer (London: Bloomsbury, 2013), 132–46.

32. David Hume, *A Treatise of Human Nature* (Glasgow: William Collins, 1962), II, III, 156.

The heart is the domain of radical subjectivity, what is most deeply and uniquely mine. On the other hand, I, as projected toward the Whole, am at a distance even from this most personal aspect of myself. I can assess my heart and, if necessary, redirect it.

These personal questions play in tandem with more general questions derived from the functioning of the notion of Being: how do we fit into the whole scheme of things? What *is* that scheme? How are we a we? Guided by these questions, one can then set to work inventorying and assessing current assumptions, theoretical and practical, open to possible transformation.

The question is how we come to terms with the way we now reflectively consider ourselves. The task here is to strive to become aware of one's proclivities and develop a kind of dialectic of discovering and purifying one's disposition, even of discovering principles for purification. There is a deep humility required here: first, allowing oneself to be measured—and that means submitting to the discipline of the tradition within which we have been raised. But further, it means allowing one's measures to be measured. While the latter might appear as arrogance—and it could be so initiated and sustained—it should involve an even deeper humility that I take to be the calling of the philosopher. But the basic machinery for doing that is exposed in what we have uncovered in this chapter. Thinking along this way, reason moves from being the slave of one's heart to being its shaper. The deepest task of reason is to form the heart, to redirect our spontaneous attractions, repulsions, and neutralities. As Hegel noted, authentic human existence is the unity of heart and head, of radical, felt, individual subjectivity and the comprehensively universal development of reason.[33]

Here deconstruction and anti-foundationalism have their

33. Hegel, *Hegel's Philosophy of Mind*, §400, 73.

place: they can be part of the deeper humility that attempts to measure our given measures. But they presuppose the nondeconstructible foundation in the structure of the field of experience that we have inventoried. They presuppose our native orientation toward the Totality that involves a profound Logos-centrism as a task approached through dialogue with various traditions.

In sum: there are the ever-present sensory fields with their differing eidetic features operative in their togetherness to serve the needs of the organism. There is the always operative functioning of the notion of Being that sets us at a distance from all determinants and thus in relation to space and time as encompassing wholes that makes abstraction possible. It is linked to the unrestrictedness of the principle of noncontradiction and to the emergence of the notion of eternity as a central aspect of our relation to the Totality. Further, the notion of Being gives us each over to ourselves as an I in relation to a Me comprising three determinants—genetic, cultural, and personal-historical—that settle into the heart as the core of one's radical subjectivity. The heart for each of us is the default mode for our self-direction. But the function of the notion of Being sets us at a distance from our hearts and poses several fundamental questions: how do we humans fit into the Whole? What is the nature of the Whole? How are we a we? And at the radical individual level: where is my heart? Is it where it ought to be? Where ought it to be? This last question joins with the first as to our place as humans in the scheme of things. The answers to these questions are already putatively settled by our cultural context and our past choices as we set about reflection on the measures of our existence. The philosophical task is not only to lay bare the overall structures involved but also to raise the basic question as to the measure of our previous measures.

4 The Cosmos Has an Inside

On the Cosmomorphic Character of Anthropos

IN HIS RECENT *Mind and Cosmos*, Thomas Nagel has turned apostate to the dogma of reductionism, which, he claims, in one or two generations will be recognized for the nonsense it is. The high priests and their acolytes have, he says, browbeaten dissenters into submission under penalty of "not being scientific."[1] Nagel's book hasn't set well with his former coreligionists, secular "true believers," especially since, though he has still preserved their second dogma—namely, atheism—he admits it is a choice and that there could be good reasons for accepting theism as compatible with the emergentist view he espouses.[2]

Nagel's strategy is, basically, to bring the scientist, thoughtfully committed to the truth of her or his discipline, into the

1. Thomas Nagel, *Mind and Cosmos: Why the Materialist Neo-Darwinian Conception of Nature Is Almost Certainly False* (Oxford: Oxford University Press, 2012), 7.

2. Ibid., *Cosmos*, 25–26, 92, 95.

evolutionary picture. That picture has emerged with modern science against the background of the Cartesian reduction of matter to measurable extension, having removed the mind of the scientist from the picture as another kind of substance, a separate mental substance, but having also reduced plants, animals, and the human body to various mechanical combinations within extension. Going beyond Cartesian methodology, modern science has developed by way of careful and enhanced visual inspection, the construction of models, and experimental manipulation. Eventually, through the development of evolutionary theory and neurophysiological observation, the scientist is brought into the picture as an evolutionary by-product whose awareness is an epiphenomenon—a bit of an embarrassment in that, though awareness is allegedly derivative and cannot originate action, it has expanded as evolution has gone forward—something John Searle has been criticizing as awfully expensive evolutionarily for something that has no function.[3] Nagel remarks that, if the way the theory of evolution has developed is correct, we should not trust the human mind in developing the theory. It is just because we trust the factual truth of evolution that we have to change the theory.[4]

Expressly referring back to thinkers like Schelling and Hegel, but also especially to Whitehead (though Nagel did not acknowledge him), and in another way Leibniz, Nagel believes that the potentialities for self-replicating, conscious, and reflectively conscious selves must be the unobserved inside for the observed outside of the scientifically unveiled Cosmos.[5] The upshot is what Nagel calls "panpsychism"—the idea that the Cosmos [is] has been pervaded by psychic powers from the very beginning,

3. John Searle, *Rationality in Action* (Cambridge, Mass.: MIT Press, 2001), 296.

4. Nagel, *Mind and Cosmos*, 27.

5. Ibid., 17.

from the Big Bang onward.[6] Such powers only show themselves under special conditions of complexity. There is no longer dead matter; there is only potentially and actually living, conscious, and rationally self-conscious matter. Since the term "matter" is so laden with history, we probably should think of another term for what abides as the lowest story in a four-story evolutionary building: cosmic potentiality in a relatively incomplex system that is, through its progressive complexification, the foundation for life, animal awareness, and rational self-awareness. It is the last level that brings the mind of the scientist into the picture. Nagel, with German Idealism, claims that rational awareness is the *telos* of the evolutionary process—as he says, in the human being, the Cosmos has become aware of itself.[7] One could say that this is an enmattering of mind, but one could equally say that it is also a re-minding of matter.

Back in the mid-twentieth century, Teilhard de Chardin, in his *Phenomenon of Man*, introduced a helpful metaphor as the distinction between what he called "the Without" of the Cosmos and what he called "the Within."[8] The first was the field explored by modern science. It was based on enhanced visual inspection of what continues to remain the exterior of the things we confront and we as so capable of being confronted. It works in terms of visually constructed models. The Within is the self-presence of human that assumes the other three levels—"matter," life, and animal awareness—each lower level subordinated to the higher, as in Aristotle's *Peri Psyches*—except that Aristotle's ontogeny for the view Nagel represents recapitulates what Aristotle explicitly rejected: evolutionary phylogeny.[9]

6. Ibid., 57–58, 61–63.

7. Ibid., 85.

8. Teilhard de Chardin, *The Phenomenon of Man*, trans. B. Wall (New York: Harper, 1959), 53ff.

9. Ibid., 138.

What I want to do in this chapter is explore the structure of this Within, beginning with the lowest, earliest levels.[10]

WHAT IS A "WITHIN," an interior? In an obvious sense, it is an inside, as the chair is inside the room and appears such through our senses. If we go into the physical interior of human beings, complex aspects are present at the macro level; but they remain a kind of exterior in relation to the many levels inside the thing available through advanced microscopy—in any case, through the extension of seeing inside the organism. Because, as Aristotle noted, vision gives us the most direct information,[11] and because its reach has been extended most spectacularly through modern scientific instrumentation, we are inclined to think it is the only kind of evidence. But no matter how deeply scientists are able to go inside a given entity visually, what remains still more interior are the potentialities of the entity in question that can in principle never be seen, for the senses present only the individual and actual, while the potentialities are universally oriented. The potentialities of any entity are another kind of interiority, always behind or beneath the visual surface, which is a kind of dashboard that we learn to manipulate without knowing what ultimately lies underneath the hood.[12] The plurality of powers in each thing has a unitary ground on the basis of which the entity in question separates itself from its environment.

10. For an extension of evolutionary theory to the expansion of the transcendentals, see Robert E. Wood, "Potentiality, Creativity, and Relationality: Creative Power as a 'New' Transcendental?," *Review of Metaphysics* 59 (December 2005): 379–401.

11. Aristotle, *On the Soul* III.437a7.

12. The image of dashboard knowledge is Owen Barfield's, in *Saving the Appearances: A Study in Idolatry* (New York: Harcourt, Brace, and World, 1957), 28–35.

In his *Sophist*, Plato has the Eleatic Stranger claim that the definition of being is the capacity of acting and being acted upon.[13] What is peculiar about those capacities is that, though each is individual in individual things, it is at the same time *universally* oriented, and its correlates are instances that are not simply individual but are individuals *of the type* correlative to the capacity. Capacities or powers require that the individuals correlated to them are instances of *kinds*. This requires that things and aspects come in natural kinds and are not simply individuals as the senses present them. So the common way of reading the relation between the universal and the individual in our experience, which places universality solely on the side of the knower, is deficient. It is already overcome in the existence of powers of acting and being acted upon and their correlates in the environment.

If to be anything is to be the locus of powers of acting and being acted upon, things require an environment of things and aspects of things correlative to the powers. Even though they are imaginatively representable as separately existing, things belong in systems. I think it goes too far to say, with Leibniz, that each thing, since it is compossible with the Whole, has that compossibility inscribed within it as the set of individual relations into which it does, can, and will enter for, among other things, that eliminates free choice.[14] But we can say that the powers of each thing are correlative to the system to which it belongs, to the sorts of things it can act upon and that can act upon it. So the underlying powers of things are within the things, but interior in relation to the visualizable exterior, no matter how refined the visualized reaches. Experimental science puts things through various tests to determine what is the specific underly-

13. Plato, *Sophist* 247.

14. Gottfried Wilhelm Leibniz, *Monadology*, §55 and §57, in *Leibniz: Monadology and Other Philosophical Writings*, edited by Robert Latta (London: Oxford University Press, 1951), 248.

ing cluster of powers within each kind of thing. This again raises the question of the unitary ground of that cluster.

When we consider development, ontogeny recapitulates phylogeny: the development of the individual recapitulates the evolutionary stages through which its type arrived on the scene. That means that the most careful and enhanced visual inspection of the fertilized human ovum yields nothing of the potentialities that lie within it and that begin to show themselves serially as the conditions for the possibility for the activation of the higher levels are realized.

EVOLUTION IS ITSELF a kind of workshop in which things are regularly but also accidentally effective and effected. It involves the random mixing of various kinds of things that, over very long periods of time and through specific kinds of conjunction, unlocks different kinds of potentialities than those exhibited by things in their hitherto regular environments. Each emergent level has its own distinctive mode of inwardness, its own nonseeable potentialities.

It is important to notice that the potentialities for life, sensory awareness, and rational awareness do not initially lie in individual things as they do in the fertilized ovum, but in the distinctive relations into which things can enter. The potentialities lie in the system. Evolution occurs as the unlocking of progressive levels of inwardness made possible by the relationality emergent through random mixing.

When self-replicating wholes finally emerge, organ-isms emerge—that is, systems of instruments for reaching the ends of a given organism. They are what I would call "proto-selves": they are self-formative, self-sustaining, self-repairing, and self-reproducing. They define themselves over against other factors in their ecosystem. Their own interaction with changing environments leads to the further emergence of types of living beings that exhibit an increasing complexity centralizing in the nervous system observable inside of the animal organism.

Go back to the earlier phases of the universe: not only are the powers of self-replication already latent, but even at the level of the atom there is a previously unexpected potentiality that unleashes an unbelievable amount of power in atomic explosions. The picturable Rutherford model of an atom showed nothing of the immense power that lies below the surface. By putting the atom through specific tests, that power was released and now is regularly unlocked through the controlled explosions in atomic power plants and atomic submarines that are mini-versions of the explosive power that constitutes the stars.

But the lowest level of immediately unlockable power present at the Big Bang is something brutal compared to the kind of refinement emergent in the evolution of life. The immense complexity and, above all, the systematic integration of consciously living beings defy the claim to chance as adequate cause. Randomness and thus contingency are not the most significant factors in evolution: their highest service is to bring into being the occasions for unlocking higher levels of complexity and integration. That kinds of things develop in these ways, involving eventually in the human organism the integration of literally billions of aspects, makes the claim to randomness alone bizarre.[15] Evolution exhibits a direction, advancing in complexity and integration, reaching a state where things not only *exist* but make an *appearance* to sensing and then to rational beings who are present to themselves. Such presence is the basis for appearance, since appearance yields an other to the being who is aware of itself, at least in the mode of feeling. This establishes a first

15. Nagel, *Mind and Cosmos*, 49.

level, the dimension of the inwardness of sensory awareness beyond the observable organism.

PHYSICISTS HAVE SPOKEN of space-time as four "uncurled" dimensions in a system that contains something like ten "curled up" dimensions. What I want to do is consider the emergence of awareness as the uncurling of distinctively "inner" dimensions, which stand as inner in relation to the outwardness of the observable, as conscious inwardness in relation to the visually available.

What is a dimension? A measurable parameter found literally in the three dimensions of space (length, breadth, and depth) and the fourth dimension of time. The things we live with and we ourselves who live with them occupy positions within space and time or, rather, space-time: we each have our own space-time line. The spatial dimensions are clear. What is important for our purposes is that here we have an instance of irreducible distinctness with inseparable conjunction. Each dimension is different than the others but cannot exist without them. Irreducible distinctness with inseparable conjunction in the structure of space-time transfers into the distinctively "interior" dimensions occupied by sensory awareness and reflective consciousness.

When I extend the metaphor of dimensions to the life of consciousness as our "inner" life, I want to make the same claim with regard to the features I will consider as I made about space and time. It is the same being who is observable in space and time whose inner dimensions are inseparably conjoined with, though essentially distinct from, the observable body to form, not two substances, a body and a mind, but a single and, I will claim, a *seven*-dimensional entity that exhibits progressively wider and deeper modes of inwardness.

Like other living beings, we too have physical boundaries:

we appear to others as spatially circumscribed. Beneath our observable surface, we, as objects of scientific investigation, present an incredible complexity, more of which is being uncovered through ever newer and more sophisticated instrumentation. One of the great scientific frontiers is the human nervous system, centered upon the brain, that governs our metabolism, our mobility, and the givens of our conscious life.

Consider the most prominent of our senses: consider vision. Let us briefly review our earlier observations. It involves light waves being reflected off of the surfaces of things present in the environment, passing through the pupils of our eyes, being refracted by the lenses and projected onto the surface of the retina where they stimulate the rods and cones that send signals, via a set of electrochemical switches along the optic nerve, to affect, finally, the visual cortex in the back of the brain. This process can be more carefully articulated through ongoing investigation, but the main lines of operation are well known. The natural conclusion is that what we see are images in the visual cortex.

This "natural" conclusion is based on what I call "the optomorphic fallacy," thinking that to be is to be object of vision, immediate or instrumentally aided. While vision leads us on practically and empirically, it systematically misleads us speculatively. The conclusion that what we see are images in the back of the brain, in the visual cortex, is no longer a scientific claim; it is a philosophic claim based upon an inference from the evidence available to visual inspection. Thus far, nothing has been said about the nature of vision itself. What is it to *see* these images?

Seeing is the locus of appearance involving a distinction and a relation between the seen and the seeing. Appearance is a completely different occurrence than what happens in and between the visually bounded bodies that make it possible. What the neurophysiologist sees is the entire process previously described within the observable boundaries of the body being examined. However, the process is not seen within the observer's head, but within the head of the observed body "out there," beyond the boundaries of the body of the observer: intentionality escapes material causality; seeing escapes the conditions that make it possible. Into the network of causal process there appears the subject-condition for appearing: the self-presence of a perceiver.

What we have is another completely different parameter, a "dimension" that is not seeable or measurable or picturable, but that belongs to the seeing, measuring, and picturing self; what I call—and, in fact what has been regularly called—an "inner" dimension, the beginning of an "interior" life, a fifth dimension emergent out of the complexity of an observable body with a highly articulated nervous system. The point of all that complexity is to furnish the basis for manifestation and self-direction. It is not that complexity that is the real subject; it is subservient to awareness, both animal and human. The peculiarity of this kind of inwardness is the ability to live in wakeful life precisely "outside" the observable four-dimensional inside, outside the spatially extended brain operating and seeable within a given perceiver. As John Dewey remarked, animals live outside their skin, cognitively *with, participating in* what is present in their environment by rising above the causal interaction that makes it possible.[16] What the scientific examination misses is precisely this peculiar inwardness that is the condition for the possibility of living outside our physical inside; and the claim that what is seen are images in the back of the brain is belied by the objects of the focal vision of the scientist that are in the environment, precisely outside his or her circumscribed body.

16. Ibid., 13.

Now the scientific account is a true account that has positive implications in what it enables ophthalmologists to do to cure problems with visual functioning. The further philosophic claim about the objects of sight being inside the brain is dictated by what seems to be an obvious assumption: that human life is contained within the observable boundaries seen every day and investigated by biologists and neurophysiologists. Seeing gives the lie to that and introduces us to a completely different kind of reality: that of appearing, of showing, of manifestation, a reality absolutely different from any seen or seeable object. All of us know by experience what it is to see, but we do not know *that* by seeing seeing; we know it by reflecting upon what it is to see visual objects, as we are now doing. The more detailed the neurophysiological account becomes, the more the differentness of ordinary seeing stands out in its strangeness.

Now, seeing is based upon a functional abstraction performed by our psychoneural system from all that is streaming in the environment between our eyes and the seen objects. Real space is filled with immediately unseeable processes such as the sound one makes in delivering a lecture, but also with all the irradiations across the electromagnetic spectrum discovered through scientific instrumentation. That functional abstraction in which we live is *phenomenally* empty space, a lived abstraction from actual, full space. This functionally concrete abstraction, phenomenal emptiness, is necessary for life, for it enables the animal to see at a distance both its prey and its predators, its mate and its offspring. An animal lives wholly within this abstractly constructed lived space moved by its appetites and has no ability to discover what underlies it. We can recognize that because we do not live wholly within phenomenal space in biological extroversion.

Of course, more than seeing is involved in the animal relation to the things present in the environment. Over time, the psychoneural system of the animal synthesizes the visual with all the other modes of sensory access so that the animal recognizes that what it sees it also hears, smells, apprehends, and tastes, driven by its appetitive needs. It learns to anticipate the typical moves of its prey and how to counter with its own moves. It does not do this consciously: its psychoneural system does it—as does ours, as well.[17] This involves a consideration of temporality.

The temporal dimension is obviously different than the spatial dimensions. At this point, I want to make what may seem a diversion: a brief consideration of time that will have a bearing on how to think about conscious inwardness. Time has three inseparable parameters—past, present, and future—and a single direction: the constant movement out of the past into the future via the present. Each parameter has an odd status, but the oddest of all is the present. The being of the past is being-no-longer, of the future, being-not-yet: as Plato noted, temporal beings are syntheses of being and non-being.[18] And as Hegel noted, they are what they are not and are not what they are.[19]

But what is the present? It's not the twenty-first century, nor 2017, nor July, nor the third, nor this afternoon, nor the hour of this reading, nor the minute of the reading of this sentence. It is not even the one-hundredth of a second that separates champion downhill skiers, nor the micro-milliseconds of the life of some subatomic particles. As we press it, we see that it has no temporal thickness: it is, like a mathematical point, temporal-

17. You should recognize in this Aristotle's description, antedating Merleau-Ponty, of the work of the *aesthesis koine* in the production of the phantasms or synthetic modes of appearing integrated with motor habits; Aristotle, *On the Soul*, 425a28ff; compare Merleau-Ponty, *Phenomenology of Perception*, trans. Donald Landes (London: Routledge, 2014), 100–105.

18. Plato, *Republic* V.478E.

19. G. W. F. Hegel, *Philosophy of Nature*, trans. A. V. Miller (Oxford: Clarendon, 2004), §258.

ly unextended, constantly moving out of the no-longer into the not-yet.

Now let us consider how an animal occupies time.[20] In sensory observation animal awareness has a certain "inner" extension in a flowing Now that gathers from the past and anticipates the future. Animal awareness not only lives outside its skin, it lives beyond the punctually flowing Now as inward temporal extension. The future has priority insofar as the point of animal awareness is to find its prey, avoid its predators, approach its mate, and care for its offspring. It learns from the past to cope with the standard moves of prey and predator. So we have the living Now, temporally extended inwardly, in which past and future are integrated.

Now consider appearance as such: it is the showing to the animal of what is other than it. Manifest otherness requires manifest sameness. It requires on the part of the animal an awareness of itself that is one with its feeling. Consider touch in this context. Whereas the other senses have localized organs, the organ for touching is the whole bodily surface. When the reticular activating system in the brain stem is, so to speak, switched on, feeling floods the body of the animal and puts it in touch with its organic base as a functional system. That is, it does not put it in touch with its liver or brain or cells, but with itself as a functional whole, so that it can move itself. The pervasive life of feeling also enables it to feel other things actively and passively with any part of its body and to move itself in response to the relation between its needs and the things appearing in its environment. Feeling is nonreflective self-awareness; appetite, pleasure, and pain are modulations of that felt self-presence. So appearance happens because a new dimension, an inward dimension of

20. The classic locus for the discussion of time is Augustine, in his *Confessions* XI.14–31.

self-presence, emerges out of the organism that enables the animal to transcend its own physical insides and to live, in an inwardly extended Now, with what is given precisely as outside. Such appearance is for the sake of animal flourishing: things appear as correspondent to animal need. Thus far the fifth, and first inward, dimension: animal awareness built upon a complex and centralized physiological inside. Animal awareness is a folding back upon the environment to establish a new relation to space and time, living outside its bodily confines, oriented toward the proximate future based upon learning from the past how to deal with what appears outside in the environment. Time is transformed from a punctual Now flowing out of the no-longer and into the not-yet, to a certain in-gathering of time in learning and anticipation.

THE ABILITY TO carry on the descriptions we have given thus far, present especially in our first consideration of spatiotemporal dimensionality, involves the operation of what Kant identified as the a priori empty forms of space and times, encompassing and directing everything we learn to fill those empty forms.[21] Thus Archbishop Usher in the middle of the seventeenth century (1650–54) filled in the otherwise empty form of time past with calculations based upon the biblical narrative. Time went all the way back to 4004 B.C.! Of course, with an immense amount of data gained through the construction of sophisticated instruments, the estimate is that the beginning of the current phase of the universe, the Big Bang, occurred some 13.75 billion years ago. And the empty spatial extensity has been filled in with estimates that parallel the time of the universe:

21. Immanuel Kant, *Critique of Pure Reason*, trans. Norman Kemp Smith (New York: St. Martin's, 1969), B 46/A 31.

astronomers now estimate the diameter of the ever-expanding universe as 91 billion light-years!

It is precisely the encompassing forms of space and time that pull us out of the living present, out of the here-and-now adjustment to what appears in the sensory environment, and put us in initially empty relation to the spatiotemporal universe as a Whole and as term of inquiry. This being outside the here and now is, as Aristotle said, a *metabasis eis allo genos*, "a leap to another genus," a deeper and more inward dimension, what I call the *sixth* dimension of intellectual awareness.[22] The operation provided by these forms of space and time allows us to abstract types from present encounter with individuals and see that they can apply to all individual instances of a given type, whenever in time and wherever in space they might appear. The recognition of types involved in animal awareness is not of the type as such but of the instance correspondent to the animal's capacities and needs. But by reason of the operation of the a priori forms of space and time, we are able to recognize a type as a type. We have the inner distance from the here and now that makes possible such ability and that supplies the "wherever" and "whenever" involved in such recognition: the concept "white" applies wherever and whenever we find it. This is what grounds language as a set of conventionally based sensory formations that hold in place our recognition of universal forms. This is what extends our relation to the extended, flowing, living Now of sensory awareness so we can deliberately go in search of the past, deliberately project our future, and thus transform our dwelling in the distinctively human living Now.

Considerations such as these led thinkers like Plotinus to claim that the body is in the soul/mind, rather than the reverse, as we are spontaneously inclined to think: *we* tend to think,

22. Aristotle, *Posterior Analytics* 1.7, and *On the Heavens* 1.1.268b1.

spontaneously and obviously, that the mind is inside the body.[23] Plotinus's claim to the reverse is not a claim to be lightly dismissed. Animals themselves live outside their observable organic boundaries. The human body is the insertion into a segment of space-time of a cosmomorphic inwardness who lives out of and beyond a relation to space-time as an encompassing Whole.

It is important to note that intellectual awareness requires the developed experience of the sensorily given and emerges out of it at some point in the maturing of the human being that is parallel to its first emergence evolutionarily. That is why intellectual awareness as a sixth dimension is necessarily conjoined with the first dimension of inwardness: sensory awareness as the fifth-dimensional ingrowth of a four-dimensional, space-time occupying organism that furnishes the organs for organism-transcending life. The "inner" life is the ability to live "out of" one's organic insides, first perceptually and appetitively in relation to the environment, then intellectually in relation to the past and the future and, indeed, in relation to the Cosmos as the horizon of all our dealings. Plotinus and, more recently, Max Scheler, in contrast to Heidegger, claim that we live *down into* time from a position that extends to the Whole and that makes possible timeless apprehensions of logic, mathematics, and value.[24]

As we noted previously, what ultimately makes intellectual operation itself possible is the notion of Being, that which, according to Aquinas and in truth, is the first notion to arise with-

23. Plotinus, *Enneads* IV.3.20.

24. Ibid., IV.8.1; Max Scheler, *Formalism in Ethics and Non-Formal Ethics of Value*, trans. Manfred Frings (Evanston, Ill.: Northwestern University Press, 1973), 385. For Scheler, though the self is a constantly accumulating process and not an underlying substance, there is a level of timeless truths that involves entry into the time of human individual existence. Nagel calls attention to the apprehension of the "eternal truths" of logic and mathematics as a significant difference in human experience; Nagel, *Mind and Cosmos*, 86.

in the mind, giving it unrestricted scope. It extends to everything and to everything about everything. And with it arises the unrestricted scope of the principle of noncontradiction, ground of all rationality. This constitutes a *seventh* and final, because all-encompassing, dimension that raises the issue of whether there might be something beyond the spatiotemporal universe itself. The deepest question is whether our being-beyond is correlate to something beyond, a Ground of being. It is on this basis that Hegel argued that *the human being* is the ontological argument, the transition from the finite to the Infinite Ground.

Thus, while animals live outside their physical inside in relation to what appears in the environment filtered out to suit animal need, we stand indeterminately beyond the here and now of animal awareness in relation to the cosmic Whole that underlies human awareness and to the possibility of even a Beyond of that. We can consequently be aware of the distinction between the surface appearance and the ontological wholeness of what appears, and we can and do direct ourselves to that wholeness in our sciences and best serve that wholeness in the love we show to others, serving them "for their own sakes."

The human intellect is radically distinguished from animal coping intelligence that is limited to operating within the environment.[25] We move from causality in the nonconscious, to participation in the sensorily conscious, and on to "the space of reasons" in distinctively human awareness.[26] Its generic difference is its reference, beyond the environment, to the Whole of what is via the emergence of the notion of Being within human awareness.

The Big Bang, the initial stage that set off the flash of light

25. See Nagel, *Mind and Cosmos*, 81. "Reason can take us beyond the appearances because it has completely general validity, rather than merely local utility."

26. John McDowell, *Having the World in View* (Cambridge, Mass.: Harvard University Press, 2009), 4–6.

setting our universe in motion over billions of years, focused a growing complexity within the limited span of living beings. In the human case, it reached incredible complexity and yet concentrated integration coordinated with opening out progressive dimensions of inwardness. The concentrated complexity and intensity in the organism are the basis for a second Big Bang: the new "inner light" that flashes forth to encompass the universe itself as its scope. Our evolutionary background is intelligible because it is the condition for the possibility of the emergence of the intellect that can activate the intelligibility of its own universal milieu through science and philosophy.

Human relation to what appears in the environment is mediated by language, by modes of meaning, by human artefacts, by institutions and practices that constitute what Searle has called "a huge invisible ontology" of first- and second-person relations, of I and You and We, and what Hegel called "Objective Spirit," or the objectification, the living on, of the projections of subjective spirits long dead as the life-world of a community.[27] One important area of such objectification is natural science as a relation to the surrounding world between practitioners living in their scientific life-world.

I WANT FINALLY to look at what it means to inhabit this seven-dimensional world, what is the concrete inwardness of the human being, human life *lived from the inside*. We observe the visually given outside of human beings, but we also observe the expressivity within the visually given of the state of mind and general comportment of those we encounter. And we come to understand better what is going on "inside" the expressive ex-

27. John Searle, *Rationality in Action*, 93; Hegel, *Hegel's Philosophy of Mind*; the second of three major sections is "Mind Objective," §§483–552.

terior by dialogue, which allows each of us to tell the other of what we have in mind, "inside" in our field of awareness and through living together that spells out that inwardness in action.

Our modes of response to what is sensorily present begin with appetitive needs that arise out of an organic substructure determined by our genetic inheritance. That sets the framework of possibilities for individual development. As we develop biologically, we grow up personally within a prestructured set of institutions and practices—that "huge invisible ontology"—mediated by our primary caregivers that opens up concrete possibilities for development. At a certain point in that development, when the notion of Being arose within us, by reason of its reference to the all-encompassing, we were set at an infinite distance from the determinations formed by the interplay of genetic endowment and cultural shaping. We came to make our own choices and thus to determine the character each of us now has. These three levels of determination—genetic, cultural, and personal-historical—provide the artist's material for us to shape in any present by the kinds of decisions we make.

What is of crucial import here is the way these three levels percolate down into the formation of what a long tradition has called "the heart," the depths of personal inwardness. In the seven-dimensional view, we are magnetically bipolar—and that in contrast to the monopolarity of the animal. The culture settles into the space between the sensorily given environment and the always exceeding Whole toward which the notion of Being gestures. The culture sets up a complex field of attractions and repulsions built up above animal need, as a variation of which we each form around ourselves our personal magnetic field. In each case, the way that field is constituted is unique to the individual. Heart is the aspect of radical subjectivity that determines how we dwell within the Whole, what is close and what far away, what attracts and repels us automatically. The favorings of one's heart furnish the default mode for the choices we make. And we might say that the purpose of intellectual and volitional development is to judge and shape one's heart as the most personal mode of inwardness, as a mode of dwelling.[28]

The heart, poised between the organic and the ontological poles, can slide into animality or rise up into mystical orientation toward the Whole, and in between it can adjust to the solicitations of the cultural milieu. It is through the heart that we dwell in the seven-dimensional world. We can remain, as Augustine sagely remarked, "outside ourselves … locked inside ourselves."[29] That is, living in sensory extroversion in function of the satisfaction of personal need, we are locked up inside our conscious selves, at a superficial level of personal existence with only a superficial relation to others.

There is a mystical mode of inwardness explored in various ways by Plotinus, Hinduism, Buddhism, and Taoism which follows out the directedness of our nature toward the Whole via the notion of Being. Heidegger gestures in that direction by his notion of *Geslassenheit* or letting-be derived from Meister Eckhart. It develops through what he calls *a sigetic*, a practice of silence that does for this inwardness what logic does for intellectual development.[30] Teresa of Avila, in her *Interior Castle*, provided a phenomenology of spiritual interiority that exhibits standard stages of development, a gradation of progressively deeper or higher inner religious states.[31] It is this kind of inwardness that sensitizes one to the depth of what appears without,

28. Dietrich von Hildebrand, chap. 8, "The Heart as the Real Self," in *The Heart: An Analysis of Human and Divine Affectivity*, ed. John Henry Crosby (South Bend, Ind.: St. Augustine's Press, 2007).

29. Augustine, *Confessions* X.xxvii.

30. Martin Heidegger, *Contributions to Philosophy: On Enowning*, trans. Parvis Emad and Kenneth Maly (Bloomington: Indiana University Press, 2000).

31. Teresa of Avila, *Interior Castle* (New York: Dover, 2007).

shown especially in Francis of Assisi, who found every natural thing to be brother or sister.[32] One is enabled to draw near to persons and things, sensitive to their spiritual depth by drawing near to their Ground.

Given the directedness of our minds to the absolute Totality that founds everything distinctively human, one can understand Augustine's religious interpretation of the directedness of our whole being toward the Whole: "You have made us for Yourself, O Lord. Our hearts are restless and shall not rest until we rest in You," for God is here understood as the Fullness of Being.[33]

SO WE ARE four-dimensional organisms, spread out in space, inwardly articulated organically, and moving through time, whose observable complexity founds the opening up of a different kind of inwardness, a self-presence through feeling as a fifth dimension whereby the animal being transcends itself to dwell outside its skin in a living Now, having learned from the past and anticipating the near future of appetitive fulfillment, able to direct itself through being in touch with its organism as a functional whole. In the human case, this fifth dimension is conjoined with and penetrated by a sixth, deeper mode of inwardness, of prereflective and reflective self-presence that allows us to transcend ourselves in various ways—intellectual and total-personal—in a relation of nearness, through the sensory appearance, to the wholeness of things and to space and time as encompassing wholes. But the ultimate ground of the human process is the notion of Being that, in flashing forth, radically transcends the environmental location of animal intelligence by

32. See Francis of Assisi, *Canticle of the Sun*, illus. Fiona French (San Francisco: Ignatius Press, 2006).

33. Augustine, *Confessions* I.1.

orienting us to the Cosmos and beyond in the question of its possible Ground. This is the seventh and deepest dimension of our being, and our deepest relation with it is with the wholeness of our being found in the heart that is restless and unable to rest until it rests in the depths of the Whole, until it dwells in relation to the Ultimate.

AS THOMAS NAGEL observed, replaying German Idealist themes, we are the Cosmos become aware of itself.[34] We are not simply aware of our immediate environment or of our cultural surround: we are aware, as the deep background of our own mode of life, of the six- and seven-dimensional inwardness of the cosmic Whole's having produced the kind of being that can display the character of that Whole and question beyond it. The Cosmos is intelligible because it is the condition for the possibility of the human intellect. We might call this view a renewed anthropomorphism; but such an anthropomorphic view of the evolutionary world is based upon the cosmomorphic character of *anthropos*, the ultimate Within of the evolutionary Cosmos; as Nagel has it, human awareness is the Cosmos become aware of itself.

A THEOLOGICAL CODA: As Aquinas noted, in a purely philosophic mode, Being is what first occurs within the mind (*in intellectu*) as the light of the mind,[35] making it "in a way, all things."[36] As infinite openness, it is "a kind of participation in the divine light."[37] But it can increase that participation only as grounded in sensory presence.

34. Nagel, *Mind and Cosmos*, 85.

35. *ST* I, q. 5, a. 2.

36. Aristotle, *On the Soul* III.8.425b27.

37. *ST* I, q. 84, a. 5.

Shifting to a purely theological mode, as the prologue to St. John's Gospel announces, the Logos is the light through whom all things were made. And that divine light filled participating light when the Logos became flesh. The Infinite entered the finite in that place where the finite is open to the Infinite, in the human being. Further, as Aquinas also claimed, from the side of the human being, though the scope of the human mind is infinite, its power is limited by its sensory groundedness. In the afterlife, the power will reach its scope through "a participation in the light of glory."[38]

As we have attempted to show in this book, the human being is cosmomorphic as the Cosmos become aware of itself. But the ultimate ground of being cosmomorphic is its being theomorphic, made in the image of God. The deepest interiority is participation in the life of God, closer to things than they are to themselves. That alone will still the restless heart.

38. Aquinas, *Summa contra Gentiles*, ch. LIII, LIV.

Bibliography

Aquinas, Thomas. *Commentary on Aristotle's "De Anima."* Translated by Kenelm Foster, Silvester Humphries, and Ralph McInerny. South Bend, Ind.: Dumb Ox, 1994.

———. *Disputed Questions on Spiritual Creatures*. Translated by Mary Fitzpatrick and John Wellmuth. Milwaukee, Wisc.: Marquette University Press, 1949.

———. *On Truth*. Translated by Robert Mulligan. 3 vols. Indianapolis: Hackett, 1994.

———. *On the Truth of the Catholic Faith: Summa contra Gentiles*. 4 vols. New York: Doubleday, 1955.

———. *Summa theologiae*. St. Louis, Mo.: Aquinas Institute, 2012.

Aristotle. *Complete Works*. Edited by Jonathan Barnes. 2 vols. Princeton: Bollinger, 1984.

———. *Metaphysics*. Translated by W. D. Ross. In *Complete Works*, vol. 2. 1984.

———. *Nicomachean Ethics*. Translated by W. D. Ross and J. Urmson. In *Complete Works*, vol. 2. 1984.

———. *On the Heavens*. Translated by J. L. Stocks. In *Complete Works*, vol. 1. 1984.

———. *On the Soul*. Translated by J. A. Smith. In *Complete Works*, vol. 1. 1984.

———. *Posterior Analytics*. Translated by Jonathan Barnes. In *Complete Works*, vol. 1. 1984.

Augustine, Aurelius. *Confessions*. Translated by William Watts. 2 vols. Cambridge, Mass.: Harvard University Press, 1977.

Barfield, Owen. *Saving the Appearances: A Study in Idolatry*. New York: Harcourt, Brace, and World, 1957.

Berkeley, George. *Three Dialogues Between Hylas and Philonous*. LaSalle, Ill.: Open Court, 1945.

Buber, Martin. *I and Thou*. Translated by Walter Kaufmann. New York: Scribners, 1972.

Churchland, Paul, and Patricia Smith Churchland. *On the Contrary*. Cambridge, Mass.: MIT Press, 1998.

Crick, Francis. *The Astonishing Hypothesis: The Scientific Search for the Soul*. New York: Simon and Schuster, 1994.

Damian, Peter. "On Divine Omnipotence." In *The Letters of Peter Damian*, translated by Owen J. Blum, vol. 4. Washington D.C.: The Catholic University of America Press, 1998.

Damisch, Hubert. *The Origin of Perspective*. Cambridge, Mass.: MIT Press, 1994.

de Chardin, Teilhard. *The Phenomenon of Man*. Translated by Bernard Wall. New York: Harper, 1959.

de La Mettrie, Julian Offrey. *Man the Machine and Man the Plant*. Translated by Richard Watson and Maya Rybalka. Indianapolis: Hackett, 1994.

Derrida, Jacques. *Of Grammatology*. Gayatri Chakravorti Spivak. Baltimore: Johns Hopkins University Press, 1976.

———. "Différance." In *Margins of Philosophy*, translated by Alan Bass, 1–27. Chicago: University of Chicago Press, 1986.

———. "The Supplement of the Copula." In *Margins of Philosophy*, translated by Alan Bass, 177–205. Chicago: University of Chicago Press, 1986.

———. *On Spirit*. Translated by Geoffrey Bennington and Rachel Bowlby. Chicago: University of Chicago Press, 1989.

de Saussure, Ferdinand. *Course on General Linguistics*. Translated by Wade Baskin. Glasgow: Fontana/Collins, 1977.

Descartes, René. *Meditations, Objections, and Replies.* Translated by Roger Ariew and Donald Cress. Indianapolis: Hackett, 2006.

Dewey, John. *Art as Experience.* New York: Putnam, 1958.

———. *Experience and Nature.* New York: Dover, 1958.

Duns Scotus, John. *Philosophical Writings.* New York: Bobbs-Merrill, 1962.

Foucault, Michael. *The Archaeology of Knowledge.* Translated by A. M. Sheridan Smith. New York: Pantheon, 1972.

Francis of Assisi. *Canticle of the Sun.* Illustrated by Fiona French. San Francisco: Ignatius Press, 2006.

Hacker, P. M. S., and Maxwell Bennett. *The Philosophical Foundations of Neuroscience.* Oxford: Blackwell, 2003.

Hacker, P. M. S., Maxwell Bennett, Daniel Dennett, and John Searle. *Neuroscience and Philosophy: Language, Mind, and Brain.* New York: Columbia University Press, 2007.

Harris, Errol. *Nature, Mind and Modern Science.* London: George Allen and Unwin, 1968.

Hegel, G. W. F. *The Science of Logic.* Translated by A. V. Miller. New York: Humanities Press, 1969.

———. *Phenomenology of Spirit.* Translated by A. V. Miller. London: Oxford University Press, 1977.

———. *Lectures on the Philosophy of Religion.* Edited and co-translated by Peter Hodgson. Berkeley: University of California Press, 1988.

———. *Encyclopaedia of Philosophical Sciences.* Vol. 1, *The Encyclopaedia Logic.* Translated by T. Gaerets, W. A. Suchting, and H. S. Harris. Indianapolis: Hackett, 1991.

———. *Lectures on the History of Philosophy.* Translated by E. S. Haldane. Lincoln: University of Nebraska Press, 1995.

———. *Philosophy of Nature.* Translated by A. V. Miller. Oxford: Clarendon, 2004.

———. *Hegel's Philosophy of Mind.* Translated by Michael Inwood. Oxford: Oxford University Press, 2007.

Heidegger, Martin. *Discourse on Thinking.* Translated by John Anderson and E. Hans Freund. New York: Harper and Row, 1966.

———. *Identity and Difference*. Translated by Joan Stambaugh. New York: Harper and Row, 1967.

———. *What Is Called Thinking?* Translated by Fred Weick and J. Glenn Gray. New York: Harper, 1968.

———. "What Are Poets For?" In *Poetry, Language, and Thought*, translated by Albert Hofstadter, 91–14. New York: Harper and Row, 1971.

———. Words." In *On the Way to Language*, translated by Peter Hertz, 139–56. New York: Harper and Row, 1971.

———. *The End of Philosophy*. Translated by Joan Stambaugh. New York: Harper and Row, 1973.

———. "Letter on Humanism." Translated by F. Capuzzi, J. Glenn Gray, and David F. Krell. In *Martin Heidegger: Basic Writings*, edited by David F. Krell, 193–242. New York: Harper and Row, 1977.

———. *The Question Concerning Technology and Other Essays*. Translated by William Lovitt. New York: Harper and Row, 1977.

———. *Being and Time*. Translated by Joan Stambaugh. Albany: SUNY Press, 1996.

———. "The Way Back into the Ground of Metaphysics." In *Pathmarks*, edited by William McNeill, 277–90. Cambridge: Cambridge University Press, 1998.

———. *Contributions to Philosophy: On Enowning*. Translated by Parvis Emad and Kenneth Maly. Bloomington: Indiana University Press, 2000.

———. *Introduction to Metaphysics*. Translated by Gregory Fried and Richard Polt. New Haven: Yale University Press, 2000.

Hume, David. *Treatise of Human Nature*. 3 vols. Glasgow: William Collins, 1962.

Husserl, Edmund. *Ideas for a Pure Phenomenology and Phenomenological Philosophy*. Translated by Daniel Dahlstrom. Indianapolis: Hackett, 2014.

Kant, Immanuel. *Critique of Pure Reason*. Translated by Norman Kemp Smith. New York: St. Martin's, 1969.

Kass, Leon. *Toward a More Natural Science*. New York: Free Press, 2008.

Kierkegaard, Søren. *Fear and Trembling*. Translated by Alasdair Hannay. London: Penguin, 1985.

Kirk, G. S., and J. E. Raven, eds. *Pre-Socratic Philosophers*. Cambridge: Cambridge University Press, 1966.

Leibniz, Gottfried Wilhelm. *Monadology*. In *Leibniz: The Monadology and Other Philosophical Writings*, edited by Robert Latta, 215–71. London: Oxford University Press, 1951.

Lonergan, Bernard. *Insight: An Essay on Human Understanding*. London: Longmans, Green, 1958.

Marcel, Gabriel. "On the Ontological Mystery." In *The Philosophy of Existentialism*, translated by Manya Harari, 9–45. New York: Citadel Press, 1961.

———. *Existential Background of Human Dignity*. Cambridge, Mass.: Harvard University Press, 1963.

McDowell, John. *Having the World in View*. Cambridge, Mass.: Harvard University Press, 2009.

Merleau-Ponty, Maurice. *Phenomenology of Perception*. Translated by Donald Landes. London: Routledge, 2014.

Nagel, Thomas. *Mind and Cosmos: Why the Materialist Neo-Darwinian Conception of Nature Is Almost Certainly False*. Oxford: Oxford University Press, 2012.

Nietzsche, Friedrich. *Thus Spoke Zarathustra: The Portable Nietzsche*. Translated by Walter Kaufmann. New York: Viking, 1954.

———. *Twilight of the Idols, and The Anti-Christ*. Translated by R. J. Hollingdale. Baltimore: Penguin, 1968.

Pieper, Josef. *The Silence of St. Thomas*. Translated by Daniel O'Connor. New York: Pantheon, 1957.

Plato. *Complete Works*. Edited by John Cooper. Indianapolis: Hackett, 1997.

———. *Parmenides*. Translated by Mary Gill and Paul Ryan. In *Complete Works*. 1997.

———. *Phaedrus*. Translated by Alexander Nehamas and Paul Woodruff. In *Complete Works*. 2005.

———. *Republic*. Translated by G. M. A. Grube and C. D. C. Reeve. In *Complete Works*. 2005.

———. *Sophist*. Translated by Nicholas White. In *Complete Works*. 2005.

———. *Theaetetus*. Translated by M. J. Levett and Myles Burnyeat. In *Complete Works*. 2005.
———. *Timaeus*. Translated by Donald Zayl. In *Complete Works*. 2005.
Plotinus. *Enneads*. Translated by Steven MacKenna. London: Penguin, 1991.
Sartre, Jean-Paul. *Being and Nothingness*. Translated by Hazel Barnes. New York: Philosophical Library, 1956.
Scheler, Max. "Ordo Amoris." In *Selected Philosophical Essays*, translated and edited by David Lachterman, 98–135. Evanston, Ill.: Northwestern University Press, 1972.
———. *Formalism in Ethics and Non-Material Ethics of Value*. Translated by Manfred Frings. Evanston, Ill.: Northwestern University Press, 1973.
Searle, John. *Rationality in Action*. Cambridge, Mass.: MIT Press, 2001.
Spinoza, Baruch. *Ethics*. Translated by R. H. M. Elwes. New York: Dover, 1955.
———. *On the Improvement of the Understanding*. Translated by R. H. M. Elwes. New York: Dover, 1955.
———. *Correspondence*. Translated by A. Wolf. Whitefish, Mont.: Kessinger, 2005.
Strasser, Stephan. *Phenomenology of Feeling: An Essay on the Phenomena of the Heart*. Translated by Robert E. Wood. Introduction by Robert E. Wood. Preface to the English edition by Paul Ricoeur. New York: Humanities Press, 1977.
Stratton, George M. "Some Preliminary Experiments on Vision without Inversion of the Retinal Image." *Psychological Review* 4 (1887): 182–87.
Strawson, Galen. *Real Materialism and Other Essays*. Oxford: Clarendon, 2008.
Teresa of Avila. *Interior Castle*. New York: Dover, 2007.
von Hildebrand, Dietrich. *The Heart: An Analysis of Human and Divine Affectivity*. Edited by J. H. Crosby. South Bend, Ind.: St. Augustine's Press, 2007.
Whitehead, Alfred North. *Process and Reality: An Essay in Cosmology*. New York: Harper, 1929.

Wood, Robert E. "Potentiality, Creativity, and Relationality: Creative Power as a 'New' Transcendental?" *Review of Metaphysics* 59 (December 2005): 379–401.

———. "On Seeing: A Phenomenological Introduction to Neuro-Psychology." *Science, Reason, and Religion: Proceedings of the American Catholic Philosophical Association Convention* (Fall 2011): 121–34.

———. "Being Human and the Question of Being: On the Unitary Ground of Individual and Cultural Pluralism." *Modern Schoolman* (Winter 2012): 53–66.

———. "Virtues, Values, and the Heart." In *Phenomenology and Virtue Ethics*, edited by Kevin Hemberg and Paul Gyllenhammer, 132–46. London: Bloomsbury, 2013.

———. First Things First: On the Priority of the Notion of Being." *Review of Metaphysics* 67, no. 4 (June 2014): 719–41.

———. "Five Bodies and a Sixth: On the Place of Awareness in an Evolutionary Universe." In *The Beautiful, the True, and the Good: Studies in the History of Thought*, 312–23. Washington, D.C.: The Catholic University of America Press, 2015.

———. "The Heart in Heidegger's Thought." *Continental Philosophy Review* (2015): 445–62.

———. "The Cosmos Has an Inside." *Existentia* 26, nos. 3–4 (2016): 465–80.

———. "The Heart: Phenomenological and Historical Studies." In manuscript.

Wooldridge, Dean. *Mechanical Man: The Physical Basis of Intelligent Life*. New York: McGraw-Hill, 1968.

Zuzuki, D. T. *Zen Buddhism: Selected Writings of D. T. Zuzuki*. Edited by William Barrett. Garden City, N.Y.: Doubleday, 1956.

Index of Names

General Index

Also from CUA Press by Robert E. Wood

The Beautiful, the True, and the Good: Studies in the History of Thought